THE LAYMAN'S GUIDE TO

CONTRACTING YOUR OWN HOME

A Step-by-Step Guide for Contracting Your Own Home

by
David Caldwell

A Publication by
Designs by Caldwell
(a division of) **425102 Alberta Ltd.**
Alberta, Canada

Published by Designs by Caldwell
Alberta, Canada

Copyright © 1994
by David J. Caldwell
5 4 3

Distributed by:

Lone Pine Publishing	Lone Pine Publishing	Lone Pine Publishing
202A - 1110 Seymour Street	206, 10426 - 81 Ave.	16149 Redmond Way, #180
Vancouver, British Columbia	Edmonton, Alberta	Redmond, Washington 98052
Canada V6B 3N3	Canada T6E 1X5	USA

Printed in Canada

First Edition: July 1994

Cataloguing in Publication Data

Caldwell, David, J. (David John), 1950 -

The Layman's Guide to Contracting Your Own Home.

Includes bibliographical references and index.
ISBN 0-9698056-0-8

1. House construction —Specifications—
Popular works.   I. Title.
TH4815..5.C35 1994 690'.837 C94-900157-0

Dedicated to:

MY WIFE

and

THE WIVES OF ALL LAY CONTRACTORS

Who must all possess the patience and understanding
to survive the many demands placed on them during
the planning, construction and move-in phases
of building your own home.

ACKNOWLEDGMENTS

To my most cherished wife
and
Mr. and Mrs. R. Goebel
and
Mr. F. Saporito
and
Mr. and Mrs. R. Moore
and
Mr. G. Maitland
for their contribution and support.

Many of the ideas and stories incorporated in this edition were taken from personal experiences which occurred during the construction of my own homes, and the homes of the above acknowledged friends and clients.
I wish to thank them for being so helpful.

Contents

SECTION 2: DESIGNING

SECTION 3: APPROVALS and ESTIMATES

SECTION 4: PROJECT MANAGEMENT

SECTION 5: APPENDICES

Quick Reference

INTRODUCTION

As housing costs increase dramatically throughout North America, the dream of owning your own home seems to be financially further away, and getting out of reach for more and more young couples. For those families who thought owning their own home was only a dream, it can now become a reality. It is the intent of this book to provide the average layman with the same advantage and knowledge held by the custom home contractors. By providing this information to potential lay builders, this book will provide the necessary knowledge to save all the money which is usually paid to the builder for supervising costs and profit. As well as saving potentially tens of thousands of dollars, being able to contract your own home will give you tremendous satisfaction and pride.

Many of the `how to´ home building books focus on the actual technical aspects of the construction. The average layman will not understand, or need to understand the how to's of wiring, heating, plumbing. However, what they do need is a practical guide for supervising a home construction project.

In this book special emphasis has been placed on presenting the material in a manner to be clearly understood by those people who have had no previous building experience. The Layman's Guide has been carefully supplemented with illustrations to clarify the text, as well as useful examples and procedures to aid in the scheduling and construction of a wood-framed house. The book does not attempt to cover all the variations of construction which may be used in different parts of the country, however where the methods vary, the basic principles and methods will still apply.

In order to successfully supervise and organize your home´s construction, the layman requires a detailed step by step resource manual to assist in selecting the subdivision, a building site, house plan type, and qualified suppliers and subcontractors capable of taking your dream from excavation to move-in. The book has been designed to enable 'the novice builder' to efficiently coordinate skilled trades in the most effective and time saving, step-by-step manner. Every element of the construction process is covered with examples, descriptions of the work needed, specifications, check lists and detailed drawings. The book also includes a glossary of building terms, checklists, a step by step construction schedule, and personal stories about the many problems that can confront you when contracting your own home.

Building a house yourself can be both a demanding and time intensive task, however, by using a common sense method of construction you will experience a new sense of confidence and gratification. You 'can' expect to build a home to fit your own personal needs and suit `your´ lifestyle.

It is hoped that this book will be a useful guide for reference purposes, and for providing basic instruction for the do-it-yourselfer, students in vocational school, building-trade apprentices, and adult extension classes.

SECTION 1: CONCEPTS

CHAPTER 1

• CHOOSING A SUBDIVISION AND NEIGHBORHOOD

When driving through subdivisions in any city or town there is always one neighborhood that stands out from the others. Sometimes it is the style of houses or simply treed boulevards. Whatever visually impresses you is a bonus as long as that visual concept remains consistent throughout the development of the subdivision.

If the developer's concept of the subdivision impresses you, make sure that there is some type of a building control in place, i.e., one that has been in place since the construction of the first house. Some developers change the controls as the development progresses, eg., change required roof lines, or reduce the house square footage requirements. For whatever reason make sure those changes would not affect your visual impression of how the neighborhood will appear upon completion.

Check the existing subdivision restrictions to reduce the potential for undesirable development of neighboring properties. Zoning restrictions are usually set by the city and developer. Make sure that all adjacent properties to your lot and the subdivision are zoned to your subdivision's benefit, and cannot be changed during the course of the area's development. Speak to a few of the neighbors to make sure that their understanding of the subdivision's architectural guidelines are the same as yours. Remember, when neighbors are having to deal with a developer who does not comply with their own guidelines, there is strength in numbers.

Figure 1-1. Streetscape showing houses with visual balance.

The character of the subdivision sometimes has a great effect on the quality, and even the future value of the subdivision and your house. A subdivision that has owner occupied, or better yet owner built homes usually means greater effort has been taken in the design and construction of the house, and therefore the subdivision's character.

You may find yourself in a position where the developer has already sold lots to builders, and the house construction is now controlled by the builders. This is a situation where you cannot totally control the construction costs, however, the architectural controls which are in place will still dictate the value and character of the subdivision. Regardless, do your research by reading the controls, checking the zoning, and speaking to your neighbors. You now have one more item to check, your builder. I will discuss this most interesting subject in greater detail later.

Community requirements of new subdivisions are usually dictated by the zoning requirements set by the city. When researching the subdivision make sure you check with the local city development officer or engineering department to confirm the future locations of schools, shopping centers, parks or playgrounds, and community centers. A good connecting road system is essential in any subdivision in order to get to work on time and access the downtown center. A dependable public transportation system through your subdivision should be in place or in the planning.

Investigate the extent of local services as this usually determines the amount to be paid as taxes for these services. Your city administration is responsible for tax assessment and the degree of service expected in return for your tax dollar. Research the tax structure for the area to make sure you can afford the taxes. Get to know the local council member living in this area, as they can usually access information more quickly than yourself. I remember it taking hours, and being transferred through the system with my question not even close to being answered. Most of you have probably experienced similar situations with red tape.

The better the local services the more attractive and valuable the property. The activities of general maintenance and safety to consider are street paving, lighting and repair, regular garbage pick up, snow removal, and winter street sanding. Is there a blue box or recycling system in place? Since a house is one of the largest lifetime investments, is it properly protected? Are the police and fire departments close by, and do they conduct a regular subdivision drive though? Is there a neighborhood watch in place? Where is the closest fire hydrant? Good street lighting and maintenance add to the efficiency of police and fire protection. Do you have local door to door mail delivery, or is the public mail box well lit and located within walking distance? Is there a public phone located close to parks? Does the subdivision have an effective public transportation system? How close in proximity are these services to the property? The last and yet most important question that you should ask is "Can you picture you and a growing family living in that neighborhood?"

Property values (present and future) are very important factors to consider since we are a more mobile society. In the past couples purchased that dream house, and planned to live in it for the rest of their lives watching the children grow, finish school, get married, and then come home for Christmas with the grandkids. During the 80's and now the 90's couples move on an average of 2 to 4 times during their lifetime. Usually affordability is the major determining factor. Most newlyweds have not had enough time to establish a savings plan, and therefore do not have enough funds for a good down payment. This usually means renting for a few years until they have saved enough, or borrowing from parents to assist with the down payment. Even with parent's assistance it is almost impossible to know which direction the family will grow, and how large a house will be required in the next few years. The first house to consider is now called a `starter home´, i.e., one that is in a young neighborhood in the process of growth and expansion. The process of building and selling, or taking a profit and purchasing or building a larger, more suitable home has now started. It is always essential to "do your research" as this will provide you with the best chance of achieving your goal of that dream house.

CHAPTER 2

• CHOOSING A LOT

If you are looking for a lot, first choose your subdivision and neighborhood as discussed earlier by satisfying yourself that you will be happy there. Look for a lot that reflects your lifestyle taken from the following points of view.

Look for a lot on a street which has minimal through traffic as this will guarantee privacy as well as safety. Key hole crescents or short cul-de-sac streets will provide you with the best lifestyle and value. Through streets with many intersections will provide you with less privacy, and are more dangerous when crossing.

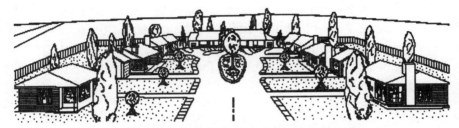

Figure 2-1. Streetscape showing key hole view.

If your subdivision is new, trees are usually non-existent and would have to be planted by the property owners. If a subdivision boulevard has been provided by the developer, the city usually requires the developer to plant at least one tree per lot, or one tree for every 20' of boulevard. The type of boulevard tree selected will dictate the size of tree or shrub type you will be selecting for your front landscaping. Mature trees are a very valuable commodity in any subdivision. They provide shade with a park-like setting, and a windbreak to prevent dust and snow from accumulating where it is not wanted.

When choosing your lot make sure that the street has been paved to the completed curb height, and that the sidewalk and curbs are clear of any cracks or chips. If this has not been done or repaired, check with the city engineer to find out when this will be done. Ask if you will be assessed for local improvement taxes, or if the developer has prepaid all local improvements. If not, you can expect to pay out several hundreds of dollars prior to or during the construction of your house for local improvements. If this is the case, a suggestion at the time of lot purchase is to consider negotiating with the developer for him to bear the cost of all the local improvement taxes.

The city hall is a must to visit prior to the purchase of the lot. The engineering or inspections department has the registered subdivision plan for this area. This plan will provide you with information about many unknown factors on, under, or beside your lot. You might discover, for instance, that there is a utility corridor at the back or side of your lot restricting the size of the house. The city might have future plans for an access walkway to a planned school playground behind or beside your lot. These factors could detract from the value and future resale of the property. The developer's sales maps might not show these potential brick walls. Find this out before, not after you have purchased your lot.

If the curbs and sidewalks are in, choose a lot that is on high ground as this will ensure better drainage away from your proposed house. Properties on lower levels usually depreciate in value because of the potential collection of ground water from adjoining higher properties as far away as the next block over. Avoid building on swampy ground, areas that are known to have a high water table, or on a lot that has large protruding boulders. This will eliminate that unknown factor which might be very costly when building your new home.

Even a seasoned builder will sometimes forget to ask the necessary questions about high water tables. As an example, a few years ago I purchased a lot from a developer in a development that appeared to be on high ground with a few houses very close to construction completion. Because of this I made the foolish assumption that everything was a go. If I had questioned the builders or owners of the houses under construction, they would have informed me that there was a high water table, and that they had to spend several thousands of additional dollars putting in extra weeping tiles on the inside of the footings. (see Figure 26-2). I found this out when I had my excavator digging the hole for the foundation. It seemed to be a very solid sand and clay base when he completed excavating, but when I returned the next day there was 24 inches of water sitting in the hole where my foundation forms were going to be installed that day.

As a result, all the subtrades and concrete suppliers had to be cancelled until further notice. The subcontractor hired to install the sanitary sewer was called in to divert the water out of the excavation site. To make a long story short, it took 7 extra hours over and above his estimate to install the required sewer pipe. I was charged extra for those 7 hours, and in addition, the construction was delayed for 5 days until the base of the excavation was dry enough to work in. All in all, because I did not bother to ask the necessary questions about the water table, it cost about 3 thousand dollars more than had been calculated. That will never happen again! (see Figure 2-2).

When driving through a desired subdivision looking at potential lots to purchase, be sure to inspect the existing rough grades of the lot as well as the property slopes of the neighboring lots. Of course there are perfect grades for every subdivision and lot, however, your house and lot budget might not allow the privilege of selecting that perfect lot. In almost all the subdivisions, the better the lots north/south orientation and water drainage the more the developer will be able to ask.

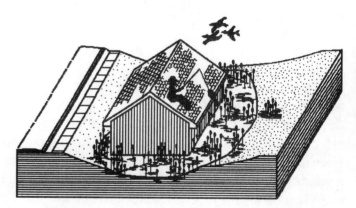

Figure 2-2. House with improper slope to drain.

There are many desirable grade slopes that will provide proper drainage away from the proposed lot. The best lot will allow any water drainage to be directed away from the house in all directions. (see Figure 2-3). This usually will apply if the subdivision and the lot are located on a burm or hill. When driving through subdivisions that are located on flat land, you will notice that there are areas of the subdivision that have been built-up with earth, and these built-up areas are usually located at the rear of the lots. This earth burm has been placed there to force the ground water or winter run-off to be directed to the front of the lot hence providing proper drainage. These back-to-front sloping lots will provide some unique and challenging house floor heights as it will be necessary for the designer to calculate the finished grade slopes of the lot in order to determine the final floor plans

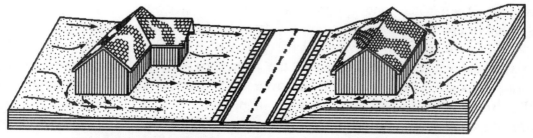

Figure 2-3. Houses showing proper slope to drain away from house.

finished floor levels. The designer will also have to consider the property around the perimeter of the house. It will also have to be built-up to allow water to travel away from the structure of the house thus providing the required drainage. (see Figure 2-3).

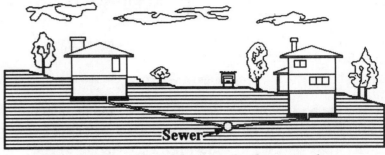

Figure 2-4. Location of the sewer line for proper slope.

Investigate the former use of the subdivision as this may directly affect your lot. If the land had been previously filled find out when this was done, if it had been compacted, and what depth of fill was brought in. If this was the case, it might be wise to get an engineer's report. He may advise a soils test at the time of excavation from which he can then determine the size of footings required to support the house structure. Before you go to this expense, talk to the neighbors beside and behind you. Did they experience any problems with their excavation, and what size of footings did they require for their house? These simple questions could save you thousands of dollars in concrete and engineering costs.

At the city engineer's office the subdivision utility plan will show the depth of the sewer and water lines, and where the connection placement might be. Make sure the sewer connection is lower than the bottom of the proposed footing. The city survey department or designated land surveyor will be able to tell you where the footing will be in conjunction to the sewer line. It will save money if the sewer and water lines can be trenched together when tying in these utilities to your house. Refer to Chapter 26 page 89 for more information.

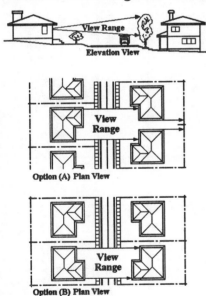

Figure 2-5. Option **(A)** showing a good view. Option **(B)** Showing a poor view.

North, south, east or west, which direction do you want your house to face? A wide rectangular lot, or a pie shaped lot with the street to the north or east is considered the most desirable as this allows the living/family area of the house to have a south/east or south/west exposure to take advantage of the sun's winter warmth. When contemplating your home design, limit any large glass areas facing north, consider shady areas created by your neighbor's house (an item usually missed), and finally, locate the direction of the winter winds. Considering the above factors could save you considerable dollars on heating bills, and reduce or eliminate potential cold drafts and condensation problems.

Check the location of telephone and cable service boxes. Are they located on your lot, and will they interfere with vehicle access to your driveway? If the driveway ices in winter would you miss hitting these flimsy metal boxes that are very expensive to replace or repair? Does the pedestal interfere with your landscaping plans, or is it possible for you to plant shrubs close by to hide it, and still allow for service access when required?

Where will the city water service valve, (the c.c.), be located, eg., in the driveway pad or front yard? It is usually marked with a flag or colored stake, and high visibility will eliminate the chance of someone hitting or driving over it. Once again this will save you hundreds of dollars in repair costs.

Know the location of your driveway and garage if not pre-determined by subdivision design controls. The location could influence the incline of the driveway. More than a 12% slope from a horizontal grade could be too much for winter access if it ices. (see Figure 2-6). Have your designer contact the surveyor

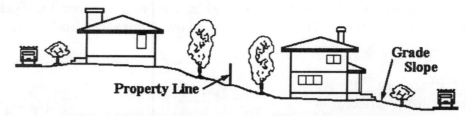

Figure 2-6. Properties showing more than a 12% grade slope.

to find the best allowable driveway slope, and then incorporate the proper slope in the working drawings. If the area chosen to build your house has a history of freezing rain, an iced driveway with too steep a slope could cause vehicles to slide when accelerating to enter or exit your garage. You should also place the driveway as close as possible to the road as allowed by the Building Code. Paving costs are directly proportional to the length of the driveway. You also have to shovel all that snow in winter! Do not reduce the workability of the lot by planning a driveway down one length to reach a garage placed at the back of the lot. If you are planning on having a wider garage make sure the lot is wide enough for your house and garage leaving the side yards required by city zoning.

Consider the subdivision design controls or streetscape of the lot in conjunction with adjoining properties. Will you be looking at their garage or RV pad? Will the design and placement of your house be compatible with your neighbors, i.e., are the garages side by side, does a porch or house protrusion block your view of down the street? Will their window look directly into one of your windows, or a spot in your backyard that you chose for a private area? These are considerations of good manners of design. (see Figure 2-5). Who ever you choose to design your home, they should be requested to do an on-site inspection of these architectural features to be sure that these problem areas will be caught prior to designing your home.

Finally the shape of your lot determines the size, visual, location and sometimes the type of house you can build. Before you commit to purchasing a lot, have an impartial appraisal made by your designer and mortgage lender. Provide them with photocopies of the building pocket of your lot, as this will provide them with all pertinent information on your lot in order to make a fair appraisal. Check with the city's tax assessment office to make sure that all property and local improvement taxes have been paid to date. Finally, to safeguard your interests and before signing on the dotted line consult a lawyer who is experienced in real estate, and knowledgeable about the developer for the subdivision. If you cover all the bases your investment is protected.

Once all your concerns are satisfied you can either purchase the lot, or sign an option to purchase on condition that you receive mortgage approval, and your house meets the development guidelines, and fits the lot. The building pocket will provide you with the last bit of information.

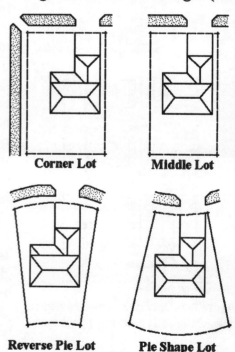

Figure 2-7. Different shapes of lots.

CHAPTER 3

• UNDERSTAND YOUR LOT

The time taken in understanding your lot warrants as much thought and care as the planning of the house. Even though the land might represent only 20 to 30 per cent of your total investment, the finished value of your house depends on how the lot and the house were planned in relation to each other. The time you take in the present planning will determine future value, livability and potential maintenance costs. All the permanent features such as driveways, trees, lawns and fences if well planned will assist in maintaining the value of your home for many years.

Visualize a building on your lot. Do not start with a preconceived idea of the house you would like to build, rather select one that will fit your chosen lot location, i.e., bungalow, 1 1/2 story or two story. Visualize it as several boxes, boxes which can be moved around to accommodate the shape and exposure of your lot. The arrangement of the house components (see Figure 16-1) must be flexible to adapt to the sun's exposure to the lot, and allow the house configuration to adjust to fit the lot. The floor plan should then naturally fall into place.

Because there are so many variables to consider, visualizing the living areas of the house as several boxes can avoid costly mistakes during the design stage of your home. The general idea when designing a house plan is to have the family areas on the south or sunny side, and to the rear away from high traffic, or directions not providing privacy. Try to orientate those areas to overlook a visually enjoyable direction, such as a garden or terraced area of the yard. Place the kitchen, laundry, and service entrance near the east and front of the house for ease of access. The stairs should be easily accessible from all areas of the house, and are best located with a northern exposure. Bedrooms and work areas service the house better if located to the north or northwest. Not every room in a house can have a perfect location, but observing a few simple rules will help avoid your worst nightmares. Remember, because a house plan works well on one side of the street that does not mean it is going to work on the other side.

If your subdivision has design controls in place some external components of your house have already been established, i.e., garage and driveway location. Either way, whether your subdivision does or does not have controls, follow these simple considerations:

1) garage and driveway accesses.
2) service locations for garbage, meter reader, milkman and newspaper deliveries.
3) access for the kids to a mud area or washroom from outside without tracking through the high maintenance areas of the house.
4) how friends and visitors will approach the house, and where the main and service doors should be located.
5) location of windows for view or security keeping in mind the direction of the summer/winter sun and the prevailing winds.

Fugure 3-1. Streetscape showing sloping properties with potential drainage problems.

Visualize the streetscape of your house, as this is the part of the house seen first by other people. First impressions of the front of the house are very important. If you have a corner lot, remember you have two streetscapes to contend with.

If the selected subdivision has varying, steep side-slopes for grades, the design of the house may force the garage to be placed under a portion of the living area. This house style is usually known as a `drive under´. The streetscape of the finished house along with its landscaping and sidewalk restrictions make this lot type the last to be developed. Because it slopes quite drastically from one side of the lot to the other, the cutting of the grass is very difficult, or the property might require retaining walls to hold back earth from falling onto the driveway or the neighboring property. The sloping property might direct the ground water onto a neighbor´s property causing basement flooding. (see Figure 3-1). In these types of subdivisions the grades will usually level out, however, the ground water distribution may be directed to the lower lying lots requiring them to add weeping tiles, sump pits, and back water check valves to stop basement flooding and sewer back-up. (see Figure 3-2).

Your front street will most likely be noisier, dirtier, and less safe than the fenced backyard of your lot. Cars and friends will frequently park on the street in the front of your house, or on your driveway causing visual obstructions. Your front view may become secondary to the more pleasing landscaped backyard, and should be windowed and landscaped with this in mind. This is why so many new homes are locating the living environment to the rear of the house, and a smaller more formal sitting room to the front. Home owners seek privacy at the back of the house with windows and doors opening the house to the outside, thus taking advantage of a more pleasing and safer view of a landscaped rock garden or terraced deck.

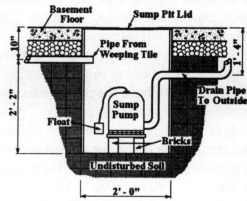

Figure 3-2. Illustration of sump pit.

NOTES

_____

_____

_____

_____

_____

_____

_____

_____

_____

_____

CHAPTER 4

• CHOOSE YOUR HOUSE STYLE

The streetscape and appearance of your house leads to discussion of the architectural style of the house.

Architectural styles, are like fashions in clothes. Original creations which are admired are usually imitated and therefore perpetuated in old and new styles such as Tudor, Cape Cod, Georgian, Western Ranch, Colonial, Californian, etc. If you have a creative designer these architectural styles can be used to capture a unique quality that will make a house stand out and look different from its neighbors.

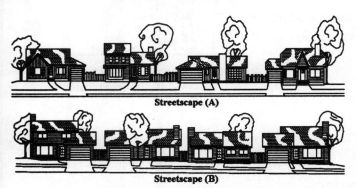

Figure 4-1. Illustration (A) showing poor streetscape. Illustration (B) showing good streetscape.

Houses on a residential street should be seen as a whole group and not as isolated buildings. The streetscape of a subdivision depends on the collective appearance of all the houses, each having common gently blending roof types, house styles, and features which benefit each neighboring house. (see Figure 4-1).

Imitating a house design is cost effective, and is done by many general contractors, especially if the shape and features of the house are visually attractive. When several reproductions of a new or existing unique style has been constructed on the same street, the effect is not one of quality but of mediocrity and monotony. Contractors who purchase many lots side by side on one street are often profit-driven to promote their best selling house plans. With usually minor changes to the colors, materials, and/or roof line they construct similar houses alternating lots and reversing the garage locations. When these houses compete with each other to attract attention the whole effect is one of confusion. This is not the best way to create a quality streetscape for the subdivision.

A subdivision assumes a pleasing and restful quality when the houses are of similar architectural styles and blend well with one another. The most functional and effective designs are those which are simple in shape and have clean lines. Therefore, when considering the streetscape of your house, think first of its general shape rather than the details. Choose a design with simple lines and proportion. For example, when considering a home's length to its height and width, these proportions can be accented with well designed entrances and window features. Remember, fads usually do not remain around very long. The resale value of your house will depend on its distinct curb appeal qualities not its similarities.

Local and regional differences exist across North America, yet the contemporary style even with its updated design changes has certain recognizable characteristics. Usually large floor to ceiling windows, and continuous or similar roof lines will combine to produce balance and symmetry of design.

Large windows are currently very fashionable, are made possible by efficient modern heating, and offer an easy transition between the indoor living environment and

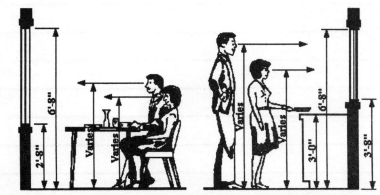

Figure 4-2. Good window sizes and locations for Dinette and Kitchen.

the outdoor leisure areas of a home. There are definite `do's` and `don'ts` to the proper location of windows. (see Figures 4-2, and 4-3).

1) Do locate the windows to take advantage of a view and the sun's radiant heat.
2) Do locate windows for security purposes away from a main street view into the house.
3) Do take advantage of garden views as this will bring the outdoors into the house.
4) Do not have glass areas facing north unless unavoidable. Then consider using triple glazed or reflective windows to keep your heating costs down.

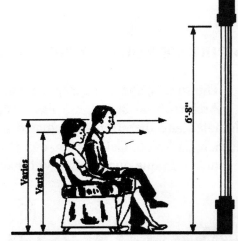

Fugure 4-3. Illustration showing good Living and family area window height.

NOTES

_____
_____
_____
_____
_____
_____
_____
_____
_____
_____
_____
_____
_____
_____

CHAPTER 5

• USING THE SUN'S HEAT

The ideal location for large glass areas is on the south or south-east side of the house. There is now a very good selection of solar-backed vertical and venetian blinds, so that windows can face south-west and west without any concerns. Window manufacturers have solar reflecting glass which has become very popular for south and west facing glassed areas, but expect to pay from 25 to 40% more on your window order for such luxuries.

Windows facing east and south get virtually the same amount of sunlight as those facing west, but during the early part of the day the sun's rays are cooler and the ultraviolet rays are not as strong, therefore, not over-heating. This is very important during the summer months of the year. At sundown the windows which face west admit the sun's rays into a room on an angle which causes overheating, also known as the greenhouse effect. (see Figure 5-1).

Figure 5-1. Location of the sun during the summer.

When designing a house locate as many windows to the east, south and west in order to assist heating the house during the winter. The resultant heating fuel savings can be as much as 30%. (see Figure 5-2). The summer greenhouse effect can be virtually eliminated with solar reflecting material on your window coverings. The extra cost paid for these materials can be paid back by the second year considering the amount of heating fuel saved during the winter.

A window's worst enemy can be its location in relation to the prevailing winter winds. Avoid placing large glass areas on sides of the house which face north or the winter winds. Unless you provide triple glazing and/or additional insulation, the windows will frost and heating bills will rise considerably. More window information is provided in Chapter 26 pages 107 - 109.

Figure 5-2. Location of the sun during the winter.

NOTES

CHAPTER 6

• FINISHING TOUCHES

Any house looks best on the lot when it appears to hug the ground. The main floor level should be brought as close as possible to the finished grade of the property. As discussed earlier, the more gradual the driveway slope the fewer the problems that will occur during a winter freeze. All this requires a careful calculation of how much excavation will be required, and how much of that material can be used for backfill to achieve a proper grade slope away from the structure. (See Figure 6-1).

It is important to achieve a proper slope or grade away from the house for good drainage. Most cities now require eaves trough downspouts to drain away from your property rather than to the storm sewer as in the past. The surveyor can calculate the depth of excavation, and the contractor who excavates the basement will usually be able to calculate (within two truck loads) how much earth he should leave to achieve the proper finished grade. This is particularly important in cases of narrow lot or zero lot line construction. In addition, it is important to know

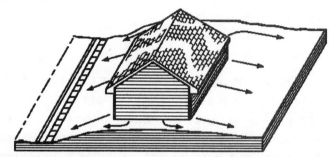

Fugure 6-1. Good grade slopes must be provided around the perimeter of the structure.

whether you want to have a standard 8 foot, or a more expensive but more functional 9 foot basement height. This choice will effect the surveyor's grade staking and soil excavation for the lot.

Cost of maintenance should be one of the first considerations when choosing your exterior materials. A well built house should require little or no maintenance during the first 7 to 10 years of its life. Exposure to weather over a period of time for any material will require some maintenance. During the life of a house this will become a major factor in the value of the property. You must therefore protect your investment by choosing durable materials which will not invite damage created by sun, moisture, and ice.

There is an excellent selection of durable, quality exterior finishes on the market. A good designer or builder can advise you as to the durability of new and existing products. Exterior finishes fall into two categories, those needing maintenance in the form of paint or stain every 5 to 8 years, and those which need little or no upkeep.

♦ Products needing maintenance:
           Cedar siding - horizontal or vertical
                    - painted wood windows
                    - plywood wood trims
           Asphalt shingles - on south/west exposures

♦ Products needing little maintenance:
           Stucco (colored) - on weathered exposures (usually only washing required)
           Aluminum,vinyl siding - on direct south/west exposures
           Pine shakes - on weathered exposures

◆   Products needing no maintenance:
          Brick or stone faces
          Stucco
          Concrete roof tiles
          Cedar shingles

For wood frame construction, most mortgage companies agree on a 30 to 35 year life expectancy of a house. Wood frame with proper maintenance will last beyond a lifetime.

Limit the number of exterior finishes used on a house. Avoid changes of materials just for the sake of variety. Depending on the style of home, other than trim features, using one finish over a large area is more pleasing to the eye.

◆   **Colours**

The use of different colors has different effects, i.e., light colors make objects look larger, and dark colors do the reverse. Bright colors used over large surfaces will make an object stand out from the surrounding surface, sometimes to the extreme. Bright colors and different textures should therefore only be used to emphasize or enhance features to which you wish to draw attention. As an example, firehalls, hospitals, and government buildings usually are constructed with brightly textured stone, brick, steel or marble entrances so they will stand out. With houses you should emphasize the window trim, gutters, soffits and fascia with white, or a color darker than the exterior wall color by about 2 color tones. This will enhance the house's features.

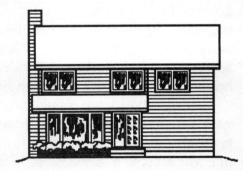

Figure 6-2. House with horizintal lines.

Color, texture or their combination can be used to fool or misdirect your eye. Horizontal lines suggest width (see Figure 6-2), and vertical lines give the impression of height. (see Figure 6-3). By using a dark color over a light color the effect of reducing the height of the lighter color will be achieved, i.e., a dark roof on a white house will connect the house to the ground. If you are planning a small house, the use of one color on the walls will give the illusion of greater size. When planning your streetscape colors there are two principal areas to consider, the roof and the walls. The colors on these two areas must not conflict, and if you are planning colored shingles or tiles this will limit your selection of color for the walls. Grey or earth tone colors on a roof will not dominate or detract from any other colors you plan on using.

Your local climate and possible selections for landscaping will also affect the selection of color and texture to be used. In areas where there are many rainy and overcast days, use white surfaces with light cheerful color as this will offset the depressed feelings created by this type of weather. Where you have or will have trees to give color and texture, simple white surfaces provide contrast. Where there are few trees the use of bright colors and different textures will make a significant difference.

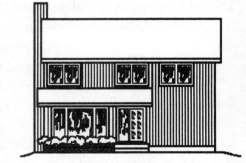

Figure 6-3. House with vertical lines.

The colors most commonly used are the simple, down-to-earth colors, i.e., greys, beiges, tans, and whites. Save those bright, dramatic, and unusual colors for your interior decorating.

CHAPTER 7

• SELECTING YOUR HOUSE TYPE

There are in reality only four basic types of house styles: bungalow, 1 1/2-story, 2-story, and split-level. There are other styles which are concepts taken from the original models, and modified to accommodate building and zoning requirements. Each has its advantages and disadvantages. It will be your choice as determined by budget, lot size, and personal preference.

Whether you are building yourself, or having a general contractor build for you, the main consideration should be "What will be the final cost?"

How does one determine which house type they can afford? One method I always use is the basic cost per square foot, but for each house type the cost per square foot differs. A simple phone call or better yet a visit to some show homes of several well established builders will provide you with that information. Once you have that information, multiply the cost per square foot times the floor area requirement of the selected house style. This should equal your budget. A builder should have the experience and knowledge to build your home within that set budget with the detailed standards and quality specifications you have discussed and included in the builder's contract. If the quote falls within your budget, and you plan to subcontract the house, you have between an 8 to 14% safety margin of play in your favor.

The bungalow has the advantage that all habitable rooms are on the same level. Because there are no stairs to climb, other than those leading to the basement, the occupant has easier access to all areas of the house. Therefore, it is less fatiguing and potentially fewer accidents occur especially where children are concerned. If your lot will allow for expansion and your intent is to make additions to your home in the future, it is often cheaper to add to a bungalow. A single level design offers a wider scope for open planning which gives the impression of spaciousness even in smaller square footages. On the other hand, this is the most expensive house type to build. A bungalow requires twice as much roof, foundation, footing and basement floor as a 2-story house of the same floor area. If not properly designed, heating and plumbing problems may be encountered if the plan is too spread out, and extra thought will be required by your installers to solve these problems Therefore, living space of more than 1,350 square foot can be provided more economically in the 1 1/2 or 2-story house.

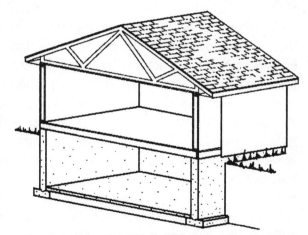

Figure 7-1. Bungalow house style.

Although this house style will not be discussed, a bi-level plan should be considered as a alternative if the bungalow costs are too high. It provides you with the main floor bungalow style, but allows you to develop the basement with less effort and cost. Because the footings of the bi-level design need to be only at the local frost level, usually 4 - 5 feet, this allows larger basement windows than most house types. The major draw back is that stairs are required to access all the main living areas, and will have to be climbed as a daily activity.

The 1 1/2-story house has many of the same features as the bungalow yet provides about 50 - 75% additional floor area under almost the same amount of roof. This means that the main floor area can contain less square footage than that of a bungalow with portions of the roof area habitable as in a 2-story.

The main floor usually contains a finished living area with a reduced yet workable bedroom area, and leaves the second floor unfinished for future development. When the need for expansion is identified the upper level can be developed to accommodate extra bedrooms and washroom facilities. The future development costs are considerably less than the main floor development as the supporting structure has already been completed. The story and a-half house style probably provides the greatest amount of livable floor area for the least initial capital outlay. Many designs for this house style require the incorporation of dormers into the initial design and construction in preparation for the future development. (See Figures 16-3 and 16-5). Dormers add substantial cost to the initial construction, but this can be offset by placing windows in the gable ends to bring in light. The insulation requirements for the 1 1/2 story house are usually more expensive because there are more wall and roof surfaces to insulate

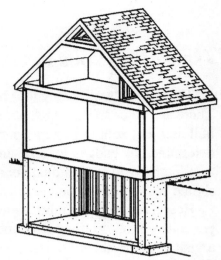

Fugure 7-2. 1 1/2 story house style.

than a bungalow or 2-story, however, the added living space this house style provides makes it a very economical design. If you will require more than three bedrooms and one bath for future development it would be best to consider a 2-story or split level house.

2-Story style: Where subdivisions have narrow lots of insufficient width to accommodate bungalows, 2-stories with similar floor layouts are often built. There are many advantages for building a 2-story. This style of house is usually more compact, and therefore easier to build. Costs are reduced as the same interior structural walls can be used for the plumbing and heating lines, and the bathrooms can be designed on top of each other sharing the same piping. Too often these small houses have a tendency to look boxy because they are about twice as high as they are wide, and this is a problem which requires careful attention from the designer. If you recall earlier, this type of difficulty can be usually overcome by emphasizing horizontal lines which will give the illusion of greater width. Another way is by co-operatively working with your neighbors on the streetscape, and grouping pairs of houses for a collective appearance.

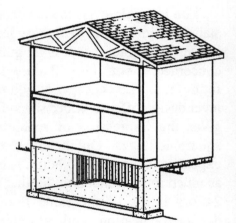

Figure 7-3. 2-story house style.

Split level style: This style of house has some of the advantages of a bungalow, and some of the 2-story style. Split levels are most useful in solving problems where there is a grade variation from one side of the lot to the other. The design provides an easier flow from floor to floor because it has only 4 to 8 steps between levels. The number of actual stairs depends on the frost level and main floor level required, and lastly whether the owners require a 3, 4, or 5 level split. Split levels provide a greater separation between the sleeping level and living level, and is similar to the 1 1/2 story home where the lower level can be left unfinished for future development. The main floor usually contains all of the living area with the bedrooms on the upper level, leaving the lower level for the furnace as well as potential family, den or laundry facilities. This level can also be used for the garage access. Many 3 level splits have crawl spaces under the living area as initially it is cheaper to excavate for a level footing, however, this level still requires the footings to reach below the frost line. I have suggested to many of my clients that the additional 3 or 4 feet of excavation costs relatively little, and can be done in a couple of extra hours when excavating the full foundation. You now have an additional level to fully finish for the future as well as a basement to place your furnace. This now makes the house more valuable for resale since it contains more square footage.

It is now time to take the surveyor's building pocket and personal sketches to see the designer you have chosen.

Before you go to the next chapter I suggest that you read chapter 23 which deals with how to choose your designer.

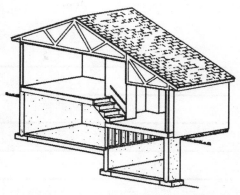

Figure 7-4. Split level house style.

NOTES

_____

_____

_____

_____

_____

_____

_____

_____

_____

_____

_____

_____

_____

_____

_____

_____

_____

NOTES

CHAPTER 8

• FINAL CONSIDERATIONS

By now you have most likely decided on a house type for you and your family. You may have your final draft presentation completed, and are ready to have your working blueprints drawn. Before you make that final decision here are a few questions that you should ask.

Take into consideration the quality or standard of living your family wants to enjoy on a day to day basis over the length of time you will be living in the house. Does the flow and layout of the house fit the way you like to live? Is it large enough now, and will it be large enough or suitable to meet your needs in the future? Will it be able to accommodate your present furniture and tools, or will you have the added expense of purchasing new furniture or storing your old? Will it be a pleasant place to live in during cold, snowy, or wet months without having to change or add devices to make the house comfortable? Will the members of your family be able to adapt themselves to the way of life that will be largely determined by the room layout of the house? There should be no doubt or compromise by the family as to the practical usability of your chosen house style. This will also guarantee its value and habitability for future owners should you decide to sell.

If you have answered yes to all questions then complete your working blueprints, but if you have any 'no's' or you 'are not too sure' re-assess the house presentation. You and your family's future well-being is at stake. Remember, you will be spending a substantial amount of money for the working blueprints, and making changes on presentations is much cheaper than re-drawing whole working drawings. Spending the extra time and money on good working drawings will save you many sleepless nights and thousands of dollars during construction.

NOTES

_____

_____

_____

_____

_____

_____

_____

_____

_____

_____

_____

NOTES

CHAPTER 9

• SEE YOUR BANKER

The best scenario to getting a house is, of course, to go to a designer, tell him what you want and what your family's needs are, have him design your dream home, and then proceed to build. Unless you have won the lottery, this is usually not the case.

Whether it is your intent to supervise and build your own house, or have one built by a general contractor, before spending time and money looking for a lot or a designer you should see your banker about pre-approved mortgages and construction financing. If not satisfied with what your bank has to offer, do some shopping around. There are many banks and trust companies willing to barter with you for your business, and sometimes at better rates. Finding out the maximum mortgage you can qualify for can place the reality of building costs into the correct perspective. The size of mortgage you will receive depends on many factors: the bank's appraised value of your proposed house (usually done by an independent appraiser), your total income and net worth, prevailing interest rates, and your available cash for a down payment and interim financing.

With the assistance of the bank's loans officer, a calculation of interim financing will be determined. These funds must be able to carry you through the lot purchase and initial construction costs prior to receiving your first mortgage advance. Of course, the more cash available for the down payment on the lot, the more likely the bank is to approve your financing.

It is common practice for a lender to require clear title on the lot prior to any construction start. Having the property in your name eliminates any claim someone else might have on the lot, and also provides the bank with its collateral. If the bank's mortgage is registered on title, and you or your contractor fail to complete the house for any reason 100% of your debt to the bank will be paid in full.

Once you have completed the applications for interim financing and mortgage approval with your selected bank be sure you understand clearly your responsibilities. The next step is to purchase the lot or take an option on the property with the conditions that you receive mortgage approval, and the house plans will fit the lot. Remember, at this point you may only have a verbal approval from your bank, and in order to purchase the lot you should have the interim financing in place or your own cash available. The following pages are typical application and loan information sheets that must be completed and approved before the final mortgage approval will be given by the bank.

Using your own savings or approved interim financing, proceed in the purchase of your lot and having the house plans completed by your designer. During the completion of the house plans you should be in contact with your mortgage officer to make sure that your mortgage application is being processed without any delays. See page 25 for sample application.

Once the working blueprints have been completed to your satisfaction, proceed to finalizing the application for mortgage. This detailed process for mortgage approval consists of providing the bank with 2 sets of blueprints in working drawing form, 2 plot plans completed by your selected surveyor, and completing the bank's building, structural and finishing specification sheets shown on pages 26 to 28 of this chapter. Once the bank has all this information, the legal mortgage documents can be completed by the bank. To start the ball rolling, the bank will require that you pay a standard application fee for an independent appraisal using your blueprints and completed cost projection sheet. (see page 24).

Many banks may not provide construction mortgages to self-builders unless you can verify that you have taken some construction courses through a local College or Post Secondary School. Upon completion of these courses you will be able to provide the bank with a letter of introduction and intent, and this should be included in your mortgage application. This letter will show the bank that you are determined in your objective to complete the construction of your home, and have the necessary knowledge of residential house construction. The following is a sample letter of introduction and intent to a Financial Institution.

Date of Letter

Name of Bank
Street Address
State/Province
Postal Code

Re: CONTRACTING HOME FOR MR. & MRS. APPLICANT

Dear Sirs:

The reason for my wife and I acting as our own general contractor are varied and numerous. From your perspective, the most important may be my experience in construction in which my father, brother and I completely renovated our family bungalow, and also added approximately 1200 square feet. All the work from mixing the cement for the piles to the finished trim was done by ourselves. It is my intent on this project to limit our contributions to organizing, scheduling, and occasional sweeping up. Through courses taken by my wife and I along with several books researched, we have developed a comprehensive three part building system.

• See Appendix for list of courses taken and list of books researched.

The first part is the seventy-five working day construction schedule which outlines the responsibilities required to keep the project moving along on time. We will be using a three day advance notice/confirmation call system to ensure punctuality from the suppliers and subtrades. As seventy-five working days translates into less than four months, I am projecting August 22 as the occupancy date (a one month allowance for delays, no shows, rain ...) as April 23 is the ground breaking day.

The second part of the system is a bound diary. This is where I keep notes of key discussions with suppliers and subtrades that took place on those dates. This will help avoid the "misunderstandings" normally associated with the subtrades and suppliers during the construction period. The diary is also used to assist in scheduling the various stages of the project.

The third and final part of the system is the supplier/subtrade index. This is where I keep copies of the signed quotations and contracts from the suppliers and subtrades.

• A minimum of three quotations per supplier/subtrade will be required.

Along with our system comes the organizational skills and dedication of my wife and I. We both have Commerce Degrees, College Business Diplomas and my wife also has an Arts degree. Obtaining these academic designations stems from our organizational ability and our determination to complete what we start.

Having both of us involved in this project is a great benefit as the potential for marital tension is lower than if only one spouse was to undertake the project. Also, not having children at this stage is certainly advantageous given the amount of evenings and weekends that will be consumed.

Probably one of our biggest reasons for acting as our own general contractor is financial. In our four and one-half years of marriage, we have done "OK" financially, comparatively speaking. It would be an understatement if I was described by my friends and family as "financially responsible". We are very aware of the cost savings in building our own home, so much so that we have already thought about building another one.

In reference to my job with _____, fortunately my work is flexible so that I can be on-site during the day when needed. I have been with _____ for almost four and one-half years now ,and my boss, Mr. _____, is aware of my intentions.

My father who has built two homes of his own and a cottage will be assisting me with some of the day to day supervision of the suppliers and subtrades. He will also be making himself available to open and lock-up the house every day for the workers.

In summary, we are confident in our ability to successfully organize the construction of our next home, and we look forward to the next opportunity to use your services again.

Sincerely,

Your Signature.
     Attached; Appendix
          Copy of quote

APPENDIX

COURSES:

Building Your Own Home, Faculty of Extension,
University of _____ , ... Fall 1991.

How to be Your Own General Contractor, Continuing Education,
_____ College, ... Jan - Mar 1992.

The Legal Aspects of Building a House, Mr. _____, Lawyer,
Continuing Education, _____ College, ... Feb 92.

BOOKS:

1) The Complete Guide to Contracting Your Home, A step by step method for home construction, by - ·Author / Publisher.
2) How to be Your Own Contractor, by - Author / Publisher.
3) Do it Yourself Contracting, Building Your Own Home, by - Author / Publisher.
4) Be Your Own Contractor, By - Author / Publisher.
5) Building Your Own Home, By - Author / Publisher.

NAME OF YOUR FINANCIAL INSTITUTION

* COST PROJECTION FOR NEW HOME CONSTRUCTION *

Branch ....

Date ....

NAME: .... APPLICATION NO. ....

LEGAL DESCRIPTION: ....

CIVIC ADDRESS: ....

BASEMENT (Forms, Footing, Concrete, Weeping Tile, Dampproofing, etc.,) .... $ _____

BASEMENT FLOOR   (Concrete and Labor, etc.).................................... $ _____

HOME PACKAGE AND/OR MATERIALS (Lumber, Nails, Siding, etc.,)............... $ _____

PLUMBING... .... $ _____

HEATING .... $ _____

ELECTRICAL WIRING AND FIXTURES (Including installations) .... $ _____

INSULATION .... $ _____

DRYWALL, TAPING, TEXTURING... $ _____

EAVES TROUGHING... $ _____

SIDING/STUCCO... $ _____

PARGING .... $ _____

PAINTING .... $ _____

KITCHEN CABINETS AND VANITIES .... $ _____

FLOOR COVERINGS .... $ _____

CERAMIC TILES, MIRRORS, TOWEL BARS, etc.,. .... $ _____

DRIVEWAY, STEPS AND SIDEWALK .... $ _____

LABOUR (give estimate of labor costs exceeding $1,000 not included above).......... $ _____

PROJECTED CONSTRUCTION COSTS $ _____

5% CONTINGENCY FACTOR $ _____

TOTAL PROJECTED CONSTRUCTION COSTS $ _____

CONFIRMED LAND VALUE $ _____

PLANS AND PERMITS $ _____

SURVEY AND ENGINEERING $ _____

LEGAL COSTS $ _____

PROJECTED TOTAL COSTS $ _____

N.B. If package home, some of the above items may be included in package
cost; if so, please indicate._____

NAME OF YOUR FINANCIAL INSTITUTION

FILE NO _____

SAMPLE - FINANCIAL STATEMENT AND LOAN REPORT   BRANCH _____ DATE _____

| [] M [] F | Last Name | | First Name | | Middle | | Age | Birth Date |

| Present Address | | City | Province | Postal Code | | Yrs. at Present Address | Phone No. |

| Previous Address (if less than 2 Years) | | Yrs. | [] Married [] Single [] Widowed
 [] Separated [] Divorced | No. of Dependents
 Excluding Spouse |

| Name and Address of Present Employer | | Yrs. | Occupation | Phone No. |

| Previous Employer and Address | | Yrs. | Occupation | Phone No. |

| Spouse's Name | Occupation | Age | Name and Address of Spouse's Employer | | Yrs. |

| Bank | Address of Branch | | Type of Account — Number |

Automobiles — Year Make		Value	Other - Trailers, Boats, Motors, Snowmobiles, etc.	Value

Real Estate Owned	Address or Legal description	Value	Mortgaged To	Amount

| Have you ever declared bankruptcy? [] Yes [] No
 If yes, are you a discharged bankrupt? [] No [] Yes — Date _____ . | Customer since:
 Date _____ |

| Social Insurance Number | Driver's License Number | Are you a Gold Card Holder
 [] Yes   [] No | Indirect Liabilities |

ASSETS	Omit Cents	LIABILITIES						
		Instalment and Charge A/C's Bank and/or Finance Co's	Date Opened	High Amount	Date of Last Payment	Monthly Payment	Balance	X If We Are To Pay
Bank a/c								
Bonds and Stocks								
Life Ins.								
Mtge. & A/S Held								
Automobiles								
Real Estate								
Other								
		Mortgages and/or Rent Under Monthly Payment						
						Sub Total		
						Surplus		
						TOTAL		

MONTHLY INCOME AND REPAYMENT (must be completed)		RELATIVES CLOSE FRIENDS Name / Address / Relationship

MONTHLY INCOME AND REPAYMENT
(must be completed)

Grose Employment Income $ _____
Spouse's Income $ _____ (%) $ _____
Other (include Family Allowance) $ _____
Total Gross Monthly Income $ _____
Less Deductions at Source $ _____
Take Home Pay $ _____
Deduct monthly payments before loan $ _____
     (See above)
Disposable Income $ _____
Total Debt Service Ratio $ _____ %

RELATIVES CLOSE FRIENDS
Name / Address / Relationship

1. _____   1. _____
2. _____   _____
3. _____   _____
4. _____   2. _____
5. _____   _____
6. _____   _____
Sub. _____   3. _____
7. _____   _____
Total _____   _____

MONTHLY BUDGET
Complete below where monthly credit obligations
exceed 30% of monthly take home pay.

Utilities	$ _____
Groceries	$ _____
Clothing	$ _____
Auto Expenses	$ _____
Insurance (Auto, Life, Fire)	$ _____
Medical and Dental	$ _____
Entertainment and Vacation	$ _____
Sundry	$ _____
Savings	$ _____
Total Monthly Expenses	$ _____
Instalment Loans Payment $ _____	
Other Monthly Payments $ _____	
Total Monthly Outlay	$ _____
Take Home Pay	$ _____
Surplus/Shortage	$ _____

The foregoing information is furnished for the purpose of obtaining advances from (_____) and
is hereby certified to be true and correct. I authorize and consent to the receipt and exchange of credit information with any credit
reporting agency, credit bureau, or any person or corporation with whom I have financial dealings, and agree that information so
received may be retained by you.

_____   _____
     WITNESS   SIGNATURE

NAME OF YOUR FINANCIAL INSTITUTION

BRANCH _____
MORTGAGE APPLICATION NO. _____

HOUSE BUILDING OUTLINE SPECIFICATION

NOTE: Starred items * Must Also Be Included On Plans.

Applicant(s):	Legal Description:	Civic Address:

EXCAVATION: Soil Type _____ Depth from Finished Grade to Footing Bearing _____

FOUNDATION: Material _____ Walls _____ Footing Size _____

CONCRETE: Type _____ Strength _____ P.S.I. at 28 days; Reinforced: [ ] Yes [ ] No

SERVICES: [ ] Mun. Water [ ] Well [ ] Mun. Sewer [ ] Septic System

[ ] Waterproofing **OR** [ ] Dampproofing Material _____

TILE DRAINS: _____ [ ] Perimeter [ ] Underfloor [ ] Sump Pump

EXTERIOR WALLS: _____ [ ] Solid Masonry [ ] Frame [ ] Other _____

EXTERIOR FINISH: Brick Type _____ [ ] Stucco [ ] Wood [ ] Other * _____

CHIMNEYS: _____ [ ] Masonry [ ] Prefabricated Size _____

FIREPLACE: * _____ [ ] Wood or Coal [ ] Electric [ ] Gas Size of Flue or Vent _____

MEMBER	SPAN	SPACING	SIZE/THICKNESS	MATERIAL AND GRADE
BEARING PARTITIONS *				
OTHER PARTITIONS *				
BASEMENT COLUMNS *				
FLOOR JOISTS *				
FLOOR JOISTS OTHER *				
CEILING JOISTS *				
ROOF RAFTERS *				
ROOF TRUSSES *				
STRUCTURAL BEAMS *				
ROOF SHEATHING				
WALL SHEATHING				
SUBFLOORING				
G. 1 S. UNDERLAYMENT				

WINDOWS: Type of Frames _____ Type of Sash _____

SPECIAL GLAZING: _____ [ ] Storm Sash [ ] Fly Screens

DOORS: Type of Frames _____ Exterior Doors - Size and Type _____

Interior Doors - Size and Type _____ Storm Doors - Size and Type _____

KITCHEN CUPBOARDS: _____ [ ] Wood [ ] Metal [ ] With Doors Counter Top Finish _____

INTERIOR FINISH: _____ [ ] Plaster DRYWALL Type _____ Thickness _____

AREA	FLOOR COVERING	WALLS AND CEILINGS	DECORATION
LIVING ROOM			
DINING ROOM			
KITCHEN			
FOYER			
BEDROOMS			
BATHROOMS			
FAMILY ROOM			
OTHER			

CONTINUED ON PAGE 27

OUTLINE SPECIFICATION SHEET CONTINUED

INSULATION: Type and Thickness: Exterior Walls _____

Basement Space: Walls _____ Ceilings_____

Roof _____ Floor _____ Slab _____

ROOFING: Type _____ Grade or weight _____ Eaves trough □ Yes □ No

PLUMBING: _____ □ 3 - Piece □ Shower over Bath □ Shower Cabinet □ Laundry Tubs □ Other

EXTRA PLUMBING: _____

DOMESTIC HOT WATER: Method of Heating _____ Capacity of Heater (Wattage) _____

Type of Tank _____ Capacity of Tank _____ Conditioner □ Yes □ No

HEATING SYSTEM: Type of Fuel _____ Type of System _____ Details _____

_____

ELECTRICAL SERVICE: No. of Amperes _____ No. of Circuits _____ Special Wiring _____

Special Equipment _____

OWNERSHIP: List of all equipment subject to Conditional Sales Contract or Rental _____

_____

SPECIAL FEATURES: _____

_____

_____

_____

TYPES OF SURFACES * Walks _____ Driveway _____ Parking Pad _____ Other _____

PARKING * _____ □ Garage - - No. of Cars _____ □ Carport □ Parking Pad

LANDSCAPING: Sodding _____ Sq. Yds. Seeding _____ Sq. Yds. □ Shrubs □ Trees

●   ●   ●   ●   ●   ●

FINISHING SPECIFICATION SHEET

PRODUCTS and MATERIALS	DETAILS	COST ALLOWANCE
1. Kitchen/Bathroom Cabinets		
2. Flooring		
3. Doors		
4. Windows and Trim		
5. Light Fixtures		
6. Appliances (Built-in)		
7. Fireplace (s): Zero Clearance or Masonry		
8. Extras		
9. Garage: Finish (Insulation/Drywall) Door Door Opener		

10. Landscaping: Concrete Drive: Double $ _____   Single $ _____

Concrete walks: $ _____

Sodding: Front $ _____   Rear $ _____

Fencing: $ _____

Deck: $ _____   Size: _____

11. Size of Home: _____ sq. ft. _____ sq. m.

I certify that the house(s) will be built in accordance with this Outline Specification and the accompanying plans and with the Residential Standards of the Building Code for your area. "Notice to Borrowers" below has been noted and will be followed.

Date	Applicant's Signature

NOTICE TO BORROWERS

♦ **THE BORROWER** is responsible for ensuring that the house is built in accordance with approved plans and specifications.

♦ No changes are to be made without the approval of (- - Your Financial Institution - -).

♦ **THE BORROWER** must ensure that construction conforms at least to standards of design and construction prescribed by the Residential Standards of The National Building Code. These Standards are available at any Government Department of Housing.

♦ One approved copy of the plans and outline specification must be avaiable on the site during construction.

♦ **FAILURE TO COMPLY WITH THE ABOVE MAY RESULT IN DELAYS OR IN THE REDUCTION OR CANCELLATION OF THE LOAN.**

♦ A homeowner or purchaser should have a written agreement with his contractor or subcontractors to ensure that the house conforms with the specifications. As this outline specification is for (- - Your Financial Institution - -) purposes only, agreements should include a more detailed specification of the proposed house.

Two copies of this outline specification and of the plans must be submitted with each application. Borrowers should include details of special or other features that may be pertinent for appraisal purposes. Additional sheets may be used where necessary. One copy will be retained by (- - Your Financial Institution - -) and one approved copy will be returned to the applicant.

CHAPTER 10

WHICH WAY TO BUILD?

• BUILDING YOURSELF

If you are planning on being your own contractor, it is most important that you do your homework. When going for a loan all the facts should be available and at your fingertips when any questions are asked. The mortgage officer may or may not grant a loan on this basis. Their final decision will be made on their estimation of your ability, financial and otherwise, to complete the construction satisfactorily. You must be able to convince them that you have the necessary knowledge and time to collect estimates and supervise the construction correctly as required by the local building code. Being your own contractor takes a lot of hard work, time, organizing ability, and a sufficient knowledge of construction not possessed by the average lay person. This is why many clients choose to pay a general contractor a fee and let them do the work.

If you are prepared for the experience of a lifetime as your own contractor you will be required to:

♦   **1)** make arrangements to purchase and have delivered at the times required the correct
     quantities of building materials needed for every step of construction.

♦   **2)** collect estimates, and select and hire subcontractors for concrete, framing, electrical,
     plumbing, heating, etc.,

♦   **3)** schedule the various stages of construction, and oversee their correct completion on
     time and on budget.

You will also be responsible for making applications for the connection of the sewer, water, power and gas lines, and determining their locations on the house and street service lines.

When collecting or reviewing estimates, check to see that the suppliers and subtrades are registered with the local Workers Compensation Board. For subtrades that are not registered or have misrepresented their position to you, make sure you understand your responsibilities by contacting the local Workers Compensation Board and getting the necessary insurance against any mishap.

Many of my clients wish to do some of the subcontract work themselves assuming that they will save on the labor costs. Before you set your mind on this undertaking, you should ask yourself a few questions. Do you have the time it takes to do the job? Will you slow down the construction schedule by doing this yourself? Will your work look professional enough? Are you able to purchase the materials at the same cost that the subcontractor can? If you also have to rent some equipment in order to do this job, how much will it add to the cost of the project? As professionals, the subcontractors will in most cases take less time to complete the same job, and that will save you some bank interest charges. As well as having their own tools they are also able to purchase the materials cheaper as they get a bulk discount from the supplier. If that is the case maybe it will be to your advantage not to take any time off work, keep bringing in a salary, and spend only your evenings and weekends inspecting and doing the service work required. You will find that the service work included in being your subcontractor's joe boy or gopher will take up much of your time, i.e., picking up hardware and lumber. You will also hear, "I forgot this.", "Could you get that?", or the best one is, "Supplying this was not in our contract". Be prepared as these are just some of the incidentals attached to being your own contractor.

If you are lucky enough to have someone in the family that is retired, has some knowledge of construction, and is willing to spend a few hours per day at the house to open it in the morning and lock it up in the evening, they can be considered a gold mine. Most all suppliers and subcontractors become more conscientious when someone who is representing the builder is around the construction site, i.e., not getting in their way, but watching them and asking the occasional question. This person will also be able to keep you informed of the progress when you are not able to be on the job site. Include this person's name and function in your letter of introduction to the bank.

NOTES

_____

_____

_____

_____

_____

_____

_____

_____

_____

_____

_____

_____

_____

_____

_____

_____

_____

_____

_____

_____

_____

CHAPTER 11

• KEEP A DIARY

I have found that a good diary or day timer is worth the cost of the house. After building my first couple of houses I found that my greatest problem was remembering what the suppliers and subtrades said, and when they said it. I use the following story to highlight the value of a diary. A finishing plumber said that he was having a problem fitting the water closet onto the subfloor, and the framer should be called back to service the work correctly before he will do anything more. In my diary I write down the date, the time the person spoke to me, and what the discussion was all about. I promptly called the framer on the phone who said he was very busy, but would be over late the next day and would call first. I recorded that conversation in my diary. The next day went by and no phone call. I called the framer the next morning, and he informed me that he was too busy and had not been near a phone, but he would be there that afternoon at 3:00 PM. That sounded like a reasonable excuse, and I let that one go, but once again I wrote all the information down in my diary. This same scenario repeated itself for several days until finally he showed, but it was 5 days later. He inspected his work and said it was ok, but the lino installer built up the subfloor incorrectly and I should call him. Once again, I wrote it all down in my diary. I called the lino installer, told him the whole story and asked when could we meet. He said tomorrow 7:30 am before he had to be on another job site. I was at the house 10 minutes early expecting the worst, but to my surprise he showed. He informed me that he had noticed the bump, but he did not say anything because he thought that maybe I was going to put in a fixture that needed the bump. I wrote down all that information.

To make a long and very frustrating story short, it was not the framer's or the lino installer's fault, but a junior rough-in plumber who installed the wrong size spacer. I was informed that they have two crews, and it was a different crew than the finishing plumbers crew that installed the spacer. The other workers forgot or did not inspect this fellow's work, and it therefore was not caught. It took over 10 days with weekends before the finishing plumber arrived and fixed the "bump". Would you believe that it was the same person who said that it was the framer's fault? He said to me, "By the way, for your information, this lump is called a 'gusset'. (see Figure 11-1). Thanking him very much, I said, "Now that you have informed me of the proper terminology, I will be able to speak to your boss on his level when I ask him to pay for the damages".

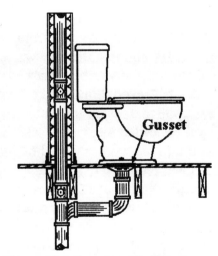

Figure 11-1. Location of toilet gusset.

My diary for those days was starting to fill up. I was really angry about the delay time as 10 days meant 10 days more interest paid to the bank. In the morning I called the plumber's boss, and told him about my delay and expenses. He informed me that he would have to speak to the plumber who worked on the job, and he would get back to me. Once more I wrote the information in the diary. Several days passed and he had not called, so I called him. He apologized, and said he had spoken to the plumber, but he could do nothing about my problem. He even had the gall to say that I should have seen it, and then called him to fix it. That was the last straw. I deducted my time and interest charges from the last payment. This is not the end of it. Over the next few months I received several invoices with interest charges which I ignored. The next month I received a final invoice along with a letter from his lawyer who I promptly contacted and explained the whole involved story. He was understanding, but requested photo copies of my diary. This discussion was also recorded in my diary. Several days later he called me, saying he had received the photo copies, had spoken with the plumber, and under the circumstances the matter would be dropped. I requested a letter from him verifying what he had just said. It was received a few days later, and the matter was closed. I am still using the same plumber, and have not had a problem with service or repair since.

This is just one example of a potentially frustrating and costly problem that can be dealt with successfully by keeping an accurate record of conversations and events. Everything done and said including names, phone numbers, discussion information, addresses and times are recorded on a day to day basis. This information has been invaluable in disputes with people who do not keep daily records. I retain my diaries for many years as a sample for my clients to view, and in the event of future problems or incidents arising. I have had to refer back to my diary on several occasions which eliminated any doubt as to which subtrade was responsible for repairing some damage.

Other suggestions include taking photographs of the construction site, purchasing a cellular phone which will provide a record of all the phone calls, and also recording in your diary the weather conditions and verbal discussions with the subcontractors. There had been a few occasions, where if I had recorded the details of a discussion or taken pictures of the site, many sleepless nights and on-site arguments with the subtrades would have been eliminated.

To again illustrate my point, in 1991 I was building a house and everything was going to work out just fine with my scheduling. The weather was cooperating, and the suppliers and subcontractors were timed perfectly. I had completed the foundation, floor joists and backfilling prior to pouring the grade beam and piles for the garage. Since the surveyors do not stake out the garage and structural piles, I requested the cribber to flag these areas with wood stakes so the water and sewer contractor would not run his pipes under any structural piles. I also marked on the foundation with paint where the water and sewer lines were located in the basement so the installer would know where to excavate. I then took a picture of the stakes and foundation markings from two views for my records. That evening I called the owner on my cellular to inform him that the site was ready, and the pile auger was scheduled for 12:00 the next morning. I also told him that my cribber marked where the garage and structural piles were located. He was pleased to hear that I had everything marked with stakes, and one of his best men would be on-site in the morning to excavate and install the water and sewer pipes. I informed him that I would not be on-site until 1:30 if his worker needed me, but he said that as long as the piles and garage were marked I would not be needed. He said that his worker would be informed, and the job would be completed before 11:00 that morning. All this was recorded in my diary. I contacted the pile contractor to inform him that all would be ready for him to drill the piles at 12:00, that my cribber would be on-site, and I had the concrete ordered for 1:30. He said everything would be taken care of. As a habit I usually do a quick drive by of the job site in the early mornings to make sure that vandals have not moved or broken anything. In addition, I always take a picture of the area(s) where someone will be working, and this morning it was the stake locations.

To make the story short, the water and sewer installer excavated where the stakes were because it was the shortest and most direct line for his pipes, so when the cribber and the pile auger arrived the cribber had to re-measure and re-stake the pile locations. The piles were drilled, and when the auger was pulled out a loud gurgling sound was heard shortly followed by water pouring out of the pile hole. The cribber who had worked for me for years knew what had happened, and he contacted me on my mobile. He suggested that they proceed with the remaining piles, as he did not think that he would hit the water or sewer line again because the earth around the next pile stake looked undisturbed. I told him to continue, then contacted the City to have the water turned off, and phoned the water/sewer installer to meet me at the job site immediately.

By the time I arrived at the construction site the city had already turned off the water at the property, and the water/sewer installer was sitting in his truck having a smoke. I suggested to him that he bring the backhoe the next day to re-excavate, repair and/or move the pipes. I also said that he should be prepared to pay for any costs or damages that he might have caused, and that those costs would be deducted from his estimate. The worker became very defensive, and said that his boss had not told him about the stakes, and that they were not near the area that he was working in. I suggested that he change his story, but when he refused I contacted his boss and described the situation to him. The owner said that the worker had been with him for 12 years, that he had no reason not to believe him, and he would not be held responsible for something I had done wrong. I informed the owner that I keep a diary of all discussions, that the cellular phone bill will show that I contacted him the night before, that the cribber will verify that he correctly marked the area with stakes, and that I had photographs of the stakes from the day before and that morning. I also had the speaker phone on so the worker and the other contractors standing near to my car could hear the whole conversation, and, if needed, they would be willing to verify that the owner was at fault for all damages and costs.

The owner quickly changed his story, suggesting that maybe his worker was trying to protect his job, that this complication had never happened before, and he would have a very serious talk with his employee. He also agreed to pay for any damages that his worker caused, and instructed me to deduct it from his estimate. If I did not keep a diary, take pictures, and purchase the cellular phone, I believe it would have been the worker's word against mine, and probably my loss. Instead, when all the damages were added up and deducted from the estimate, I owed the water/sewer installer $ 76.74. What does that tell you?

♦ **List of things to do:**

It is also very helpful if you keep a daily list of 'things to do' that is separate from your diary. This will provide you with a record and daily check list which should contain the date, phone calls to be made, items to be picked up, and people to be seen. Some days you will find that you were only able to complete the important items on your list. Any of the items missed during that day should go onto the list for the next day. Always make your list in the evenings after all your work and phone calls have been completed, and your mind and body have had some time to relax. This list will help organize your day making the construction schedule more workable and easier to complete.

In summary, these suggestions provide a historical record to remain with the house, and some interesting and funny stories for your friends.

NOTES

CHAPTER 12

• HIRING A GENERAL CONTRACTOR

If you have decided to build your house using a contractor to supervise the construction, here are a few pointers.

When selecting a contractor, it is best to see his work. If he is a small builder get at least 3 references of clients who will allow you to view their homes. If you wish to work with a larger builder, they usually have show homes for you to walk through. This however is not always the best method to assure consistent quality since a show home usually represents their best work. As in the first case it is fair to ask for three references who have been living in a builder-constructed home for about two years. This will verify that the builder constructs a quality home, and provides appropriate post-construction maintenance service.

◆ BUYER BEWARE

A contractor who is an efficient operator with a low overhead will save you money. I believe the contractor driving a truck and on-site working with his subcontractors will most likely give you good value. On the other hand, the contractor who drives a mercedes or cadillac, and spends most of his time at the office or the racket ball court should trigger concerns about where your money is going, and who is supervising the construction.

It is not necessary to be an expert in the construction industry to come to an informed opinion on how well a house is constructed. Even a professional inspector cannot see what is behind drywall or under carpets. Like you, the inspector can only judge what he can see, and then form an impression. The clues are there, eg., the care in which the finishing has been completed. Check the hand railings, door trims, baseboards and kitchen cabinets to see if they have been sanded smooth, and the joints and corners tight or properly sealed with caulking. If the show home is decorated and finished with frills, or you get a car or something free, you can guess that there is something wrong. A builder who is conscientious about providing a high quality product does not need gimmicks. You will plainly see quality in the plumbing appliances, finishing hardware, heating equipment and kitchen cabinets if they are well built and intended to stand the test of time.

As discussed in chapter 4 another way of checking for a quality product is to see if the builder is using up to date plans. Is he using a stock plan that has been imitated or repeated over and over again throughout the subdivision? Then question if the builder's primary interest is more in making a quick return on his money than providing a new, interesting and quality product to the potential home owner.

◆ BUILDER CONTRACTS

There should always be a detailed written contract with the builder that includes the following: an agreed price and method of payment which corresponds with your bank's construction advances, a specified period for completion, working drawings and detailed specifications, rights and responsibilities of both parties, terms of warranties, method of settlement for incomplete or unsatisfactory work, responsibility for fire and liability insurance, coverage for Workers Compensation, protection of property, and any other details you wish to have added. Once this is completed you will need the services of a lawyer to review the contract. It is sometimes best to choose a lawyer who has dealt with your bank, and is knowledgeable about construction mortgages in order to protect your interests, not the builders.

The contract should specify in detail the work to be done by the builder, but no contract is perfect. Sometimes there are surprises that will affect the final construction cost. A common but good example is the designer does not show driveway, sidewalk, sidewalk/patio/driveway piles, and concrete patio locations, the builder, supplier or subcontractor might not include them in his costs. This error could cost you thousands of dollars. The contractor will leave those items out, which means that his estimate would look cheaper, and he will be able to charge extra for these items because they have not been included in the blueprint and therefore his contract. He's got you !!!

If you are building with a contractor the bank will require a copy of the building contract, builder's specifications, and the completed bank's specification sheets along with the blueprints, plot plans and financial application as previously discussed. Your building contractor or designer will be able to help you complete these documents. Do not let anyone but the bank see your financial package, as that is personal information. Provide the bank with your building contract unsigned, as they might require changes; make sure your mortgage loan is approved before you sign the building contract.

You may wish to use the contractor's design services or use one of their in-stock house plans, or choose your own designer to draw a custom plan specific to your needs. Even if you choose their services, treat the design as if it is to be a custom plan. Having your own design provides the builder with your detailed specifications, and does not hold you to their design specifications. Once again, make sure your house plans have not been compromised. Spending the extra time and money on good working drawings will eliminate many misunderstandings, and save you sleepless nights and potentially thousands of dollars during construction.

Once you have completed all of the above, and your bank has approved the builder and the mortgage package, you may start the construction. Some financial institutions will allow you to start construction of the foundation up to backfilling stage before mortgage approval provided that you own the property and have a valid foundation permit from the city engineering department. >>> This practice is not very good. If the bank or city inspector requires structural changes to the foundation after reviewing the floor plans you might be inviting extra expenses. The worst scenario will require demolition of the existing foundation in order to make the structural changes. There goes your budget and possibly your dream home.

Through no fault of yours or the builders, sometimes materials or products cannot be delivered on schedule, or become out-of-stock items, eg., windows, kitchen cabinets, floor coverings and finishing products. You must be prepared to choose a new product which usually means upgrading and purchasing a more expensive line. When walking through the house during the framing, you might feel that a room is too small or you require a wall removed, a window added, an opening enlarged, or a fireplace added. If this happens the contractor will most likely extra bill you for the service. Even if you feel this will not happen, have a backup plan with your lender by applying for a pre-approved personal loan. 5% of the construction cost is a safe figure. If not used, you owe nothing. During my years of construction experience this backup has come in handy more than once.

It is necessary for the home owner to know the standard of work and quality of materials that will be going into their house. Most builders have a list of standard construction specifications that they provide to their clients for each house. Many builders have different specifications for different categories of houses such as standard contract, executive style, and custom built. The quality of materials and selections available for each of these house types is determined by the final construction cost, i.e., the more custom the house the greater the selection and cost.

This specification sheet is also valuable if you plan on building the house yourself. It provides the suppliers and subcontractors with a detailed list of expected standards required by you and your family.

SAMPLE CONSTRUCTION SPECIFICATIONS
****

CONSTRUCTION SCHEDULE "A"
- Footings to rest on undisturbed soil;
- 2500 psi (17.5 mpa) concrete type 10 in steel reinforced concrete walls and footings;
- 2000 psi (15.0 mpa), no air, type 10 basement concrete floor and 6 mil poly vapor barrier over tamped, sand-fill base to top of footing;
- Asphalt damp-proofing on exterior and interior concrete walls;
- Continuous 4" diameter weeping tile covered with minimum 6" crushed rock around house draining to a sewer, drainage ditch or dry well;
- Minimum grade of lumber to be # 2 spruce or better - all fir to be # 2 or better;
- 2 x 6 exterior walls with R20 fiberglass insulation;
- R40 ceiling insulation in main house attic, R20 ceiling insulation in garage, R32 in areas where the upper floor of the house extends over the garage area;
- 6 mil poly vapor barrier caulked at all joints;
- Poly vapor hats around electrical outlets on all outside walls and cold ceilings;
- Header joists to be doubled when exceeding 4'-0" in length;
- Trimmer joists to be doubled when header joists exceed 32";
- 2 x 10 fir floor joists @ 12" o.c. with 2 x 2 cross-bridging and strapping;
- Provide bridging at 7'-0" on center or less;
- Double joists under all non-load bearing partitions;
- Hollow steel adjustable columns to be 2 7/8" diameter with 4"x4" steel top and bottom plates - top plates to be bolted into structural beams with 2" lag bolts;
- Structural wood columns to be a minimum 6"x6";
- 3/4" Sturdywood or plywood tongue and groove subfloor, glued and screwed;
- 2 x 4 @ 16" on center frost walls in basement with R20 insulation, full height;
- Roof trusses are 24" on center engineered trusses;
- 1/2" Spruce plywood roof sheathing with pine or cedar shakes - slopes of 4/12 or greater;
- Asphalt roof shingles #210 - slopes of 4/12 or greater.
                    #235 - slopes of less than 4/12;
- 3/8" Sturdywood or plywood wall sheathing with siding - 1/2" with stucco;
- 1/2" drywall throughout and all cold ceilings 5/8" rigid fireboard;
- 3000 psi (20.7 mpa) concrete, exposed aggregate for all walkways and driveway, all reinforced with 1/2" rebar, 20" x 20" grid.
- Stairs: - rise 8" maximum - run 9" - tread width 9 1/4" minimum
- Headroom clearance 6'-4" minimum - stair width 2'-10" minimum;
- Stairs with more than 2 risers to have a handrail 2'-6" (minimum) above nosing;
- 2" x 10" fir deck header board attached to house with building paper, flashing and caulking.

EXTERIOR SCHEDULE "B"
- California style stucco, siding as per plan;
- Prefinished metal soffits, fascia and eaves troughs;
- Double-glazed wood casement window units, as per plan;
- Building paper around all windows, doors and exterior areas where water seepage can occur;
- Caulking around all windows, doors and exterior areas where water seepage can occur;

- Insulated steel exterior doors, as per plan;
- Frost-free water taps, location as per plan;
- Exterior weatherproof electrical outlets, locations as per plan;
- Lot is rough-graded, slope to drain away from building;
- Deck and piles as per plan;

GARAGE SCHEDULE "C"
- Garage floor (where applicable) - 3000 psi (20.7 mpa), no air, type 10 concrete over sand-fill base, c/w wire mesh and rebar (20" x 20" grid);
- Drive through 13' x 14' concrete garage pad - 3000 psi (20.7 mpa) type 10 concrete over sand-fill base, c/w rebar (20" x 20" grid);
- Attached garage to have reinforced foundation and grade beam, as per plan, with piles or footings, frost void form under grade beam;
- Plug-ins for electrical garage door openers;
- 220 volt power plug in garage;
- 1 - 16' x 7' and 2 - 9' x 7' metal sectional overhead doors, stained/painted/insulated;
- 5/8" drywall fireboard on all common walls;
- 1/2" drywall and insulated garage (R 20 ceilings and R 12 walls);
- 1/2 H.P. garage door openers with 2 remote controls;
- Frost-free water tap;
- Garage floor drain with check valve to eliminate back up.

INTERIOR SPECIFICATIONS SCHEDULE "D"
Plumbing as follows:
- Double stainless steel kitchen sink, c/w sink garbage disposal and vegetable spray;
- Two porcelain waterclosets;
- American standard bidet or equivalent;
- Custom tile with waterproof tile board in master bath shower unit , c/w glass shower door enclosure (modesty lines);
- Fiberglass one piece tub and shower units in main baths.
- Fiberglass whirlpool, c/w 6 jets, 72" x 36";
- Porcelain basins in all bathrooms;
- Ice-maker water line connection for kitchen refrigerator;
- Single-lever brass taps throughout;
- Insulated # 50 (40 imp. gallon) hot water tank;
- Roughed in plumbing for 4 piece bath in basement;

GENERAL SPECIFICATIONS SCHEDULE "E":
- Complete vacuum system with power head;
- Complete built-in dishwasher;
- Complete built-in trash compactor;
- Complete built-in microwave;
- 220 outlets for cook top, oven and dryer;
- 100 amp. electrical service with copper wiring;
- 2 - 100,000 BTU mid-efficient furnaces;
- Smoke detectors as per building code;

- Dryer vent to outside;
- Mirrors over vanity sinks, 36" or by vanity width;
- Weiser or equivalent- black pearl handles in all bathrooms, bedrooms and master bedroom, etc.,
- Weiser or equivalent (locking) - black pearl handles in all bathrooms and master bedroom doors;
- Bathroom outlets on ground fault interrupter circuit, as per Electrical Code;
- Ceramic tiles 18" over whirlpool and decking; full height above kitchen countertops; splash row above all vanities, one row around built-in shower unit;
- Paint grade modular doors (colonist style or equivalent), pine casings and baseboards throughout, mirrored closet doors in master bedroom and foyer;
- Garage stair access to basement of house with railings;
- Textured spray ceilings throughout;
- Glass french glass doors in dining room, den, master bath;
- TV outlets and telephone outlets, as required by owner;
- Oak spindle railings and handrails painted to match decor;
- Gas fireplace or "Ø" clearance metal fireplaces, c/w log lighter - glass & brass with remote lighter;
- Custom kitchen, bathroom and laundry cabinets, as per plan by cabinet maker;
- Complete security system with glass break, motion sensors, door pads and window pads;
- Complete intercom system with bathroom speakers;
- 40 oz. stain protected carpets with 3/8" commercial foam underlay throughout halls, stairs, bedrooms and living areas;
- Candide or equivalent lino in kitchen, mud area, laundry room and main bathroom;
- Ceramic tile in front entry, master bath ensuite;
- Cedar deck - approx. 15' x 26' with railings and steps, 5' x 6' concrete pads 3000 psi (20.7 mpa) type 10 concrete over sand fill base, c/w rebar (20" x 20" grid) to be placed at bottom of deck steps.

NOTES

_____

_____

_____

_____

_____

_____

_____

_____

_____

_____

NOTES

_____

_____

_____

_____

_____

_____

_____

_____

_____

_____

_____

_____

_____

_____

_____

_____

_____

_____

_____

_____

_____

_____

_____

_____

_____

CHAPTER 13

● ALTERNATE CONSTRUCTION CONTRACTS

You probably purchased this book in order to gain more knowledge of housing construction. If you are also interested in saving some money, and are willing to spend some time finding out about construction costs, there are alternate contracts available with some builders. The building contract and specification details remain the same whichever way you choose to build, i.e., the information for mortgage purposes as well as any of the protection clauses.

◆ 1) COST PLUS:

This method favored by some clients allows home owners to select the quality and price of materials and products resulting in a known final cost, plus a 10% surcharge of the construction value. The contractor in this instance is guaranteed a profit at the end of construction. The building contract and specification details remain the same as stated in the contract information for mortgage purposes as well as any of the protection clauses, i.e., the work to be done by the builder, a date for completion, the plans and specifications to be followed and completed in a good and workman like manner, the price to be paid, and the times at which payments are to be made.

The guaranteed profit allows the builder to declare a profit margin up front so that the client does not have to worry about any hidden or unforeseen costs. This, however does not let the contractor off the hook. They still must provide you with a total cost for construction, and the client is able to review all estimates received by the contractor, and reject or approve those quotes. To this end include in the contract that the contractor supply you with a minimum of two estimates for each different areas of construction, i.e., electrical, framing, plumbing, windows, heating, etc.

In order to further protect yourself, it would be prudent to collect estimates of your own for comparison with the contractors. The estimate should provide you with a detailed list of required materials and labor included in the price. Also remember, estimates should have a validation date that gives sufficient time for the house to be completed. In preparation for this task the different areas of construction will be described in greater detail later in the book.

◆ 2) PROJECT MANAGEMENT:

This method of construction is the happy medium between the 'do it yourselfer', and hiring a general contractor. As potential home owners become more knowledgeable about construction, the contractors become more innovative in methods of attracting clients. The concept of project management came about when the market became oversaturated with builders. Some of these builders, usually the smaller ones, were unable to survive, and they returned to the work force sometimes working for other builders as supervisors or contracting out their experience and services to owner-builders.

The project manager's job description does not require him to do the bookwork. They are contracted to provide specified services to the builder, i.e., the home owner. In return for his service he can request construction advances or a lump sum payment upon completion of the home. These services include the application for permits, gathering of estimates, ordering materials, scheduling subcontractors for all stages of construction, ordering of materials and supervising the construction on a day-to-day basis. The project manager's building contract and specification details remain the same as stated previously in chapter 12 as well as the inclusion of any of the protection clauses.

Once again, it is a good idea to collect estimates of your own for comparison with the project managers.

Remember as a 'builder' when collecting or reviewing estimates to check that the suppliers and subtrades are registered with their local Workers Compensation Board, and if they are not to find out the extent of your responsibilities.

In this business arrangement it is now the responsibility of the home owner to keep a proper accounting of the mortgage advances and payments to the suppliers and subtrades. When the project manager provides you with the invoices, be sure the lien hold back percentage on the payments is the same percentage the solicitor has held back from the mortgage advances. You will remain responsible for making applications for the connection of sewer, water, power and gas; the project manager will mark the correct locations for the house and street service lines. You will also be required to provide adequate theft and liability insurance on the property and house.

Regardless of whether you choose to build or have your home built, if you review chapter 26 and find that a certain step has not been delegated, before signing any contract clarify who is responsible for its completion. Clients have told me, the biggest complaint they have with constructing a house is dealing with the subtrades. Complaints that the last subtrade has not completed his work correctly or cleaned up after himself will require a judgment call on whether or not to call the subtrade back, do it yourself, or have someone else fix it and charge that cost back to the subtrade's final payment. In all instances it is wise to inform the tradesman and give him a chance to do the repairs, or to tell him of the probable consequences.

NOTES

_____

_____

_____

_____

_____

_____

_____

_____

_____

_____

_____

_____

_____

_____

_____

CHAPTER 14

● BUILDER'S LIEN ACT

The building contract should be drafted to protect you against any claims under the Builder's Lien Act of the state/province in which you are building. The Builder's Lien Act may vary depending on the Local or Federal requirements. In all cases workers, suppliers of materials, subcontractors and contractors who have provided materials or labor on the construction of your home are given the right to lien your property. If they have not been paid, the Act may give them the right to sell the land and to hold back sufficient funds for their unpaid accounts. If your contractor neglects to pay his workers from the draws, you may personally be required to pay them, even though you have already paid the full advance to the contractor. Make sure the contract protects you, and spells out what makes the contractor personally liable for payment. Your contract with the builder should also include provisions for the holdback of funds on each draw and payment until all lien rights on the property have expired, and proof in writing is provided that the builder has paid all outstanding accounts to the date of the last mortgage draw. The amount of the holdback will vary in different states/provinces, but is usually at least 15%. In some cases the mortgage contract with the bank may require the solicitor to retain this holdback from each draw as protection. This is not required in all states/provinces, however, it is a good practice to ask the bank what their policy is, and confirm that the solicitor be directed not to advance any monies to the contractor until you are assured your interests are protected.

NOTES

_____

_____

_____

_____

_____

_____

_____

_____

_____

_____

_____

_____

_____

NOTES

CHAPTER 15

● INSPECTIONS

As the construction progresses, inspections will be required by the City and your Mortgage holder. The purpose of the inspections is to ensure that the construction conforms with the approved plans and specifications, and local building standards. These inspections must be done as the lender requires information on the progress of construction so the solicitor can be advised of dates and amounts of money to be advanced. At least three inspections will be required by the City; your lender might require more. If you require more inspections, or if the inspector felt that you had not completed all the required work and suggests an additional inspection, be prepared to pay a fee for the additional inspection.

Here is a sample of the stages at which the inspections are usually required:

INSPECTION NO 1
Basement ready for backfill

☐ EXCAVATION   ☐ FOUNDATION
☐ BACKFILL   ☐ FRAMING, SHEATHING
☐ WATERPROOFING   ☐ SUBFLOOR
☐ WEEPING TILE

INSPECTION NO 2
Interior ready for drywall

☐ ROOF COMPLETE   ☐ ROUGH PLUMBING
☐ ROUGH WIRING   ☐ FRAMING, SHEATHING
☐ EXTERIOR DOORS, WINDOWS
☐ ROUGHED-IN HEATING
☐ BASEMENT FLOOR POURED
☐ INSULATION, VAPOR BARRIER

INSPECTION NO 3
House ready for painting/finishing

☐ DRYWALL AND TAPING COMPLETE
☐ HEATING EQUIPMENT INSTALLED

INSPECTION NO 4
House complete - Ready for occupancy

☐ INTERIOR DOORS HUNG ☐ FLOORS FINISHED
☐ EXTERIOR COMPLETE
☐ KITCHEN CUPBOARDS INSTALLED
☐ PLUMBING COMPLETE   ☐ ELECTRICAL COMPLETE
   (fixtures installed)   (fixtures installed)
☐ SITE IMPROVEMENTS   ☐ BASEBOARDS, TRIM, etc.

Advances post-inspection can be arranged with some flexibility to satisfy both parties. You may request the advances be paid directly to yourself, and you in turn pay the contractor according to the payment schedule in the contract.

When the final inspection is requested, your home should be completed and ready for occupancy, and at this time the lender will advance the final payment. The solicitor usually deducts his fees for services from this amount, and forwards the remainder to the home owner or builder less the required lien hold-back. This holdback is held in a trust account until the lien period passes, usually about 45 days.

● MORTGAGE PAYMENTS

The lender will inform you that there is interim interest due on the mortgage advances made to you. It is calculated from the date the first advance was made to the date the first actual mortgage payment is due, usually the first of a specific month. The interest rate charged for this period is based on the mortgage rate negotiated between you and the bank, and is usually much better than a personal loan rate.

The solicitor at this time will be instructed to register the title to the property in your name, and identify the lender as mortgage holder on the property. Monthly mortgage payments of principal and interest are now due. You may be required to pay monthly property taxes to the lender, or choose to pay them yourself as monthly payments, or a lump sum payment to the city when the taxes are due.

NOTES

_____

_____

_____

_____

_____

_____

_____

_____

_____

_____

_____

_____

_____

_____

_____

_____

_____

_____

_____

_____

_____

_____

_____

CHAPTER 16

• CREATING YOUR HOUSE PLAN

Even before looking for the subdivision and lot, most people already have a minds-eye picture of the appearance, style and basic interior floor layout of their 'special' home. This chapter will assist you in understanding the basis component elements which creates a floor plan.

There are potentially five elements required in a house: formal/casual living, sleeping, family, working and garage areas. These must be arranged in relation to each other in order to satisfy the different living habits of the family, and take maximum advantage of the house location on the lot.

A successful floor plan will be created from the orientation of all the component areas which impact on the various activities of each individual member of the family group. In order to satisfy the family members you may have to consider alternate groupings of the elements before you find the most workable arrangement, regardless of having already decided on a bungalow, bi-level, 1 1/2 story, split level or 2 story house. Keep in mind, when placing the four elements of living, sleeping, family and work areas that connecting them with a minimal amount of hallway will allow a smooth transition from one area to another without passing 'through' any of the elements.(see Figure 16-1).

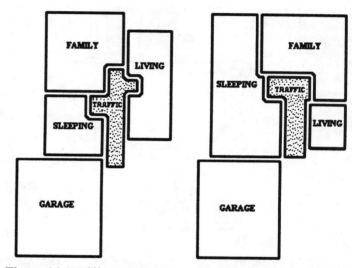

Figure 16-1. Different orientations for elements of a floor plan.

As your house plan evolves and begins to function, you will always have to correlate room size to the potential arrangement of furniture. Consideration must be given to negotiating around furniture without cramping the space, or constantly bumping into the other family members. However, avoid focusing solely on furniture 'fit' during the design phase as furniture can always be moved, changed or stored, but your family life style and budget are less flexible.

To illustrate my point there are 2 very basic principles to remember in the design phase. A square has less wall area than any other shape generally used in house planning. The greater the wall area the higher the construction and maintenance costs. A larger outside wall perimeter will result in increased weather exposure, and therefore increased fuel consumpion during cooler winter months. A more simple wall shape = less enclosing wall surface = cheaper construction costs. (see Figure 16-2).

The more corners you have in your design the more expensive, whether they are inside or outside. Setbacks, courtyards or protrusions of any kind on a house weather differently, and sometimes cause

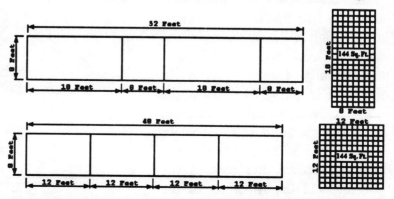

Figure16-2. Illustration showing a square and a rectangular wall with the same floor square footage. The square being more economical.

wind swirls, snow drifts or overheating when exposed to the prevailing winds or the summer sun.

Unfortunately, the square box style of house is less attractive than the rectangular style with a longer frontage or streetscape. This is especially true of the 2-story salt box styles. Some compromise between visual impressions and costs will have to be made in your design.

The simpler the roof, the lower the cost. Those roof types with a variety of slopes and changes in roof surface add to the cost of the construction, i.e., dormers, hips, bay and garden windows as well as porches require additional cutting, framing and flashing. Varying slopes and angles on the more complicated roof types will require more maintenance and add to long term expense. (see Figure 16-3).

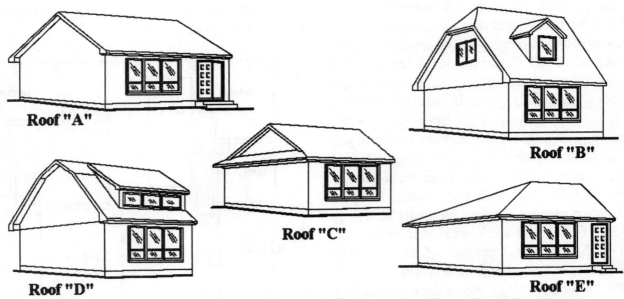

Figure 16-3. • (A) - Gable Roof.   • (B) - Clerestory Gable with a Bullnose and a Gable Dormer.
• (C) - Dutch Gable Roof.   • (D) - Gambrel Roof with Shed Dormer Roof.   • (E) - Hip Roof.

Most builders in the 90's are using premanufactured roof trusses which can vary from a high pitch 8/12 roof to a low pitch 2/12 roof. Generally there is not a great difference between cost and efficiency of 2/12 to 5/12 roof slopes, however, when the slope exceeds the 5/12 slope costs increase drastically. The slope of a roof is the distance the roof rises over a standard 12" length, eg., a rise of 4" to a typical length of 12" equals a 4/12 roof slope.(see Figure 16-4).

Increased costs need to be considered when finishing the exteriors of high pitch gable ends and dormer windows, as it takes substantially more material to finish the larger surface area of the walls as well as the overhangs. Even though the dormer windows will provide for more living space in those rooms, the cost of the dormer, and the finishing of the added floor space will have to be carefully weighed with the in-

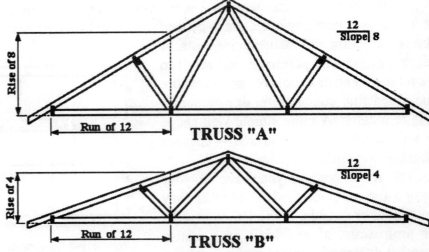

Figure 16-4. Roof slope types. • (A) - 8/12 truss slope. • (B) - 4/12 truss slope.

creased cost to build the dormer. Can you justify the added expense? (see Figure 16-5).

As discussed in chapter 7, it is possible with the steeper, pitched gable roof to use the space under the roof slope for living or storage space, eg., the 1 1/2 story design. This is not possible with a hipped roof which is also more expensive due to the more complicated truss design, additional labor of laying the shingles, flashings and roof capping at the intersections, and the additional eaves trough lengths.

In summary, if you have a limited construction budget consider very carefully the basic design of your house. The exercise of visualizing your house as several boxes can avoid costly mistakes during the design stage of your home.

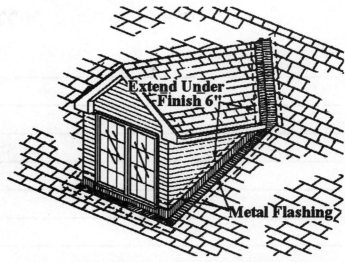

Fugure 16-5. Illustration of a Gable Dormer window.

NOTES

_____

_____

_____

_____

_____

_____

_____

_____

_____

_____

_____

_____

_____

_____

_____

_____

NOTES

CHAPTER 17

• SPACE PLANNING

♦ LIVING ROOM

We are all beginning to wonder whether the living room still has a viable place in the house plan. Many clients have it for convenience of resale, a space to display their antique or good furniture, or an area into which the dining room table can be expanded when the family gathers for special family or group functions. I believe the living room should be called the collecting room where a family member might wander to reflect on the past or contemplate the future. A room where valued family portraits, collectibles and fine works of art are displayed. A room simple in shape, functional in design, warm and pleasing to the eye.

♦ DINING ROOM

A separate, formal dining area clearly defined from the living room is desirable. Its definition can be created with a sofa, lamps, door entry or even window coverings. This room should be large enough to accommodate a table with at least 6 chairs, a buffet and china cabinet. It should have sufficient circulation space around the chairs when they are pulled out even with the edge of the table. For convenience, all the cutlery, silver, table linens, glassware and other accessories should be stored in the dining room or within a short carrying distance. A direct and unobstructed access from the kitchen to the dining room is essential for easy meal preparation and service. In fact, a minimal amount of effort should be required for the whole sequence of cooking, setting the table, serving, eating, clearing up, washing up and storing thus producing a choreography of uncomplicated events.

♦ FAMILY ROOM

Of all the rooms in a house, this family area has to be the most adaptable because of its many different uses. As the gathering place for family and friends as a whole, or where each individual family member seeks their own interests, it has to combine the requirements for family life including entertainment, relaxation, study, and children's activities.

The family area is often the focal point of the home. Informal, open and easily accessed from the other areas of the house, it is common and worthwhile to combine the other connecting rooms in order to visually expand this central living space. By combining the kitchen and breakfast nook or dinette, a very large country style environment is created even in a small house. Its length and width may vary. A long narrow room is harder to furnish than a square room which is more useful and complementary for different furniture placements. Be sure that the design allows for sufficient unbroken wall area for furniture placement, and that walls are not cut up for door openings. The number, height and width of windows also affects the amount of wall space for furniture. If this room can be located at the back of the house for privacy, the windows and doors opening the house to the

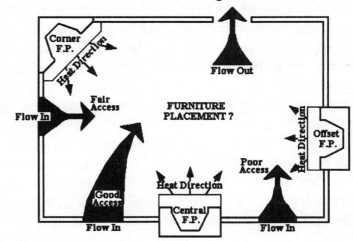

Fugure 17-1. Illustration showing traffic flow around fireplaces.

outside will take advantage of the more pleasing and safer view of a landscaped rock garden or terraced deck. (See Figure 29-3).

When planning a fireplace location, place it away from the normal traffic route, visible from all corners of the room, and preferably on an inside wall. This will not obstruct an outside view, or permit the conduction of cold air from the outside.(see Figure 17-2). Remember that a fireplace will take up a length of 5 to 6 feet, and it is a good idea to have at least 3'-6" of wall length on either side of the fireplace hearth so that furniture can be comfortably grouped around the fireplace. Locating the fireplace close to a doorway or traffic pattern will complicate furniture placement. Consider the hearth as part of the fireplace especially if it will be raised, and allow sufficient space for traffic circulation and furniture placement. (see Figure 17-1).

A major consideration when designing your floor plan is the location of the front, back, and side entrance doors. It is undesirable to have an access door opening in the direction of the family area, kitchen, or dining areas because of cold draughts, loss of privacy, and clean-up considerations.

Locate the family area so that it cannot be used as a primary passage from one area

Figure 17-2. Picture of potential drafts in living areas.

of the house to the other. If this is not possible accept only a small corner or area of intrusion.

Make sure that you have sufficient electrical outlets around the room. Think of locating several plugs and the cable TV outlet where your entertainment center might be located. Consider placement of vacuum outlets, intercom units, telephone and stereo speakers in relation to each other when planning the room and potential furniture arrangements.

♦ INFORMAL DINING

With lifestyles changing so drastically during the 80's and 90's, the breakfast nook or dinette areas are being designed as part of the family room separated by railings or half walls. This informal eating area so close to the kitchen does not entirely duplicate the function of a main dining room.

With the mother and father usually working, the family unit requires an area allowing visual and verbal communication while the day to day activities of cooking, playing and relaxing are taking place close by. Without this country kitchen atmosphere the family unit could become segregated to their own chosen areas with little social interaction.

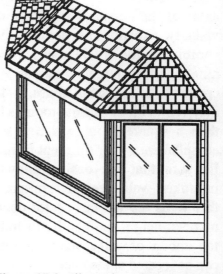

In order to add space to a potentially small breakfast nook, many home builders take advantage of a bay window addition. A bay window will visually enlarge the eating area, and provide additional space for tables and chairs. It will also allow a broader view of the backyard play areas allowing adults to keep an eye on the children's activities. Bay windows also allow more sunlight into the nook, kitchen and family rooms of the house as a result of the additional

Figure 17-3. Illustration of bay window.

glass placed into the three angled walls of the bay compared to the smaller glass area of a flat wall. (see Figure 17-3).

The family activities that occur around the dinette or nook should be discussed in great detail as the needs of the family will determine the use and size of this area. Does your family gather for regular meals three times a day, or do the family members use this area as a stop-over on the freeway to the refrigerator?

◆ KITCHEN

The kitchen is almost certainly the most widely used, and probably the most expensive room in your house. A kitchen that is both attractive and efficient is every chef and homemaker's ideal. But perfect kitchens do not just 'happen'. They are the result of a well-thought-out plan that considers everything from the size of the family to the location of overhead lighting, and many other important details in between.

Questions to consider include how often do I entertain? Is the entertaining that I do considered formal or informal? Is there more than one cook in the family, and do they work together in the preparation of the meals? How modern do I want the appearance, and how much cleaning do I want to do? Do I design it for resale, or will we be living in it long enough to benefit from some extra frills?

A self-appraisal by the cook or cooks is the next step. Is cooking a happy experience, or is it a necessary evil? Will it be necessary to have the kitchen work area clean all the time, or do open shelves and hanging utensils create a working environment that you find pleasing? Take a good look at your present kitchen. What do you like, or not like about its appearance and efficiency? Which appliances do you consider indispensable in your daily work schedule? Have you ever considered a built-in oven, or a cook top with a barbecue in an island?

The three primary work areas in any kitchen center are the range, refrigerator, and sink. Most of the steps you take when preparing a meal will be from one of these work centers to another. The normal work flow pattern is from food storage (refrigerator), to preparation (sink and adjoining space), to cooking (range), to final serving preparations. After meals, the pattern flows from the clean-up area (sink) to the storage (refrigerator).

The first principle in kitchen design is to place these three primary work centers in a convenient relationship to each other. The most satisfactory way of achieving this is to locate them so that their arrangement forms what is commonly called the 'work triangle'. Within reason, the shorter the distance around the perimeter of the triangle, the fewer the steps you will have to take to prepare meals and clean-up afterwards. Of all the work centers in the kitchen, ovens are the least active, therefore built-in ovens and microwaves are often located outside the work triangle.

The triangle may be scaled to suit the actual floor area allocated for your kitchen, but ideally the sum of the sides should not be less than 12 feet and no more than 22 feet. The shape of the working triangle may be modified according to the available space, and your personal requirements. The most compact, equal-sided triangle occurs with the U-shaped kitchen; the most elongated is with the L-shaped or island kitchen. Ideally, and for maximum efficiency, the traffic flow into the kitchen from adjoining rooms, or from other activity centers in the kitchen should not cross the working triangle.

Ensure that there is sufficient design information on the blueprints to relate your exact requirements to the kitchen salesperson. It is the designer's responsibility to show the locations of all the fixtures such as the range, refrigerator, dishwasher, cook tops, microwaves and built-ins that might effect the flow or traffic pattern, and hence the workability of the kitchen. Islands and special storage or pantry areas should be drawn or notations made on the working drawings sufficient for the cabinet supplier and installer to understand. This detailed information is required by the cabinet designer to determine the placement and locations of his cabinets in accordance with the owner's appliance requirements.

NOTES

CHAPTER 18

● BASIC KITCHEN LAYOUTS

Five designs stand out in residential kitchens today. Each of the kitchen floor layouts present both advantages and disadvantages, and all have many variations. For a large family, choose a design that will allow good circulation, eg., the U-shaped or L-shaped kitchen which can expand to become rounded or octagonal. The design selected should suite your budget and family lifestyle.

1) *Two Wall or Corridor Kitchen* is an efficient floor plan that provides maximum counter and cabinet space in a long, narrow room. It eliminates hard-to-reach corner space. The working triangle is compact and efficient with major appliances strategically located on two facing walls. For comfort and safety, the corridor should have a minimum of 4 feet between the counter tops of the base cabinets to minimize congestion when two people are working. Wall to wall width should be no more than 10 feet and no less than 8 feet. However functional, this kitchen is seldom used in residential applications.

2) *L-Shaped Kitchen* arranges appliances and work space along two perpendicular counters (see Figure 18-2). This is the most popular kitchen plan because it adapts to a wide variety of arrangements.

Frequently one run or end of the L doubles as a room divider. This plan frees floor space for other uses, and directs traffic away from the cook's work area. With this design, placing the sink and appliances can be challenging. Modifications of the L shape include incorporation of a work center, pantry or telephone desk.

If not thought out or designed properly, this kitchen plan can result in the appliances being located at the ends of the L, or located too close together with no counter separation. If the sink, refrigerator, and cooktop are too far apart, the work triangle will be exhausting.

3) *U-Shaped Kitchen* requires more space, but is considered by many experts to be the most efficient plan because of its compact work triangle, and the easy separation of the work area from family traffic patterns.

This floor plan divides appliances and work areas among three connected counters arranged in a U shape, with some or all parts of the U extending into the room without wall support, similar to that of the L shaped kitchen. Generally, the

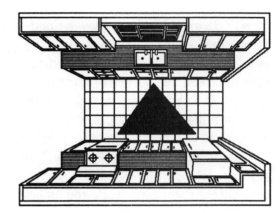

Figure 18-1. Corridor kitchen.

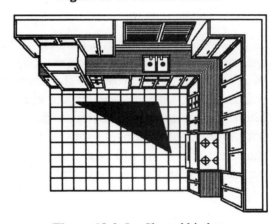

Figure 18-2. L - Shaped kitchen.

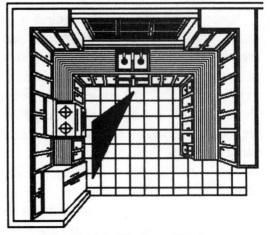

Figure 18-3. U - Shaped kitchen.

sink is placed at the base of the U, and the range/cooktop and refrigerator on the adjacent sides. The result is a tight work triangle that eliminates wasted effort. Counter space is continuous, and ample storage is made available. Problems will arise if the kitchen is narrower than 6 feet between the two sides. Special angled corner cabinets with lazy-susans are usually required in this kitchen shape to fully utilize storage space.

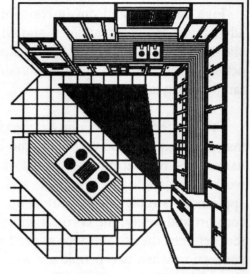

4) *Island Kitchen* has helped expand and create more interesting styles of kitchen floor layouts in recent years. The island is usually centrally located and free standing; the island can be fixed or portable adding extra work space wherever needed. If fixed, it may contain a sink or cooktop. An island can also provide an eating counter or coffee bar. Because of this adaptability, islands can effectively control traffic, provide a tighter working triangle, create more work space, and add storage. Islands are especially useful in dividing large kitchen spaces into functional work and entertainment areas.

Figure 18-4. Island Kitchen.

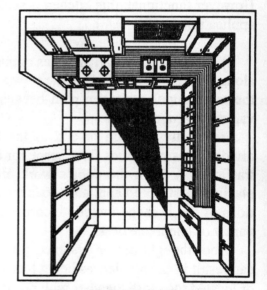

5) *Peninsula Kitchen* is another variation of the U-shaped or L-shaped plan. In this arrangement, one side of the U or L is used as a peninsula that provides the convenience of more storage space, and effectively divides the kitchen from the dining area. The upper and/or lower cabinets may be made accessible from both sides. The peninsula may also function as a breakfast or snack bar. There should be a minimum of 6 feet between the side work areas with a minimum of 5' between facing work centers.

Figure 18-5. Penninsula kitchen.

♦ **KITCHEN CENTERS**
   Since each kitchen task requires a particular working surface and/or appliance, as well as specialized tools and basic ingredients, it makes sense to organize 'centers' by locating the needed equipment, space and appliances in a convenient work configuration.
   1) *The Sink Center* handles food washing, trimming, dish clean-up, and garbage disposal. Appliances include the sink, a mechanical garbage disposer, a dishwasher, and sometimes a trash compactor. Counters are recommended on both sides of the sink. A dishwasher is commonly placed to the left of the sink; a left-handed person, though, may prefer to have it on the right side. Provide storage for soap and other cleaning materials and towels near the sink. Locate dish and flatware storage close to the dishwasher for ease in unloading.
   2) *The Refrigerator Center* handles the storage of the perishable foods. A refrigerator/freezer combination, or a separate freezer unit allows longer storage of many foods with easy accessibility when frozen products are required. An infrequently used freezer though, may be best located in an adjacent utility room,

or in the garage freeing valuable kitchen space. A counter on the door latch side of the refrigerator serves as a handling area for groceries on their way to and from the refrigerator.

Theoretically, the refrigerator works by dissipating heat, and the range works to create heat, therefore, it is inefficient and unwise to put the refrigerator next to the range. When the two are in close contact, heat from the range may interfere with the refrigerator's cooling system.

3) *The Preparation Center* is the place for mixing and preparing foods prior to cooking or serving. A major requirement is a counter surface for cutting and chopping which may be either built-in or easily accessible. One or more electric outlet for a food processor, mixer, and other small appliances used in food preparation is handy. Sometimes a lower-than-usual counter is installed as a 'baking' center.

Ample storage in close proximity for dry ingredients, baking pans, casseroles, and utensils needed for measuring and mixing is helpful. Safe knife storage is a must. Storage of small, infrequently used appliances such as the food processor, juicer, mixer and electric can opener presents a challenge. Where counter space is limited an appliance center or pull-out pantry might be the answer. The best place for this center is near the refrigerator and/or sink for access of food products and equipment clean-up.

4) *The Cooking Center*, once the location of the sink and refrigerator has been been decided, is usually self-evident. Separate wall ovens and microwave ovens can be placed anywhere outside the working triangle that is convenient. Remember to allow adequate heat-resistant counter space next to the range. Ranges and some cooking surfaces require overhead exhaust hoods and ventilation fans to filter grease-laden air, and remove cooking odors. Some cooktop units vent through the floor. Duct work for exhaust systems can be concealed inside the above-range cabinet and vented through the roof, or installed in the lower cabinet and vented through the floor.

For safety, it is wise to position a gas range away from a window as drafts can extinguish the pilot light or flame, and curtains blowing in the breeze could catch fire.

5) *Counter space* is the remaining space available after the major appliances within the work triangle have determined the basic layout of the kitchen. Whenever the layout permits, counter surfaces should be provided on both sides of the sink, refrigerator and range. The working surfaces are the same height throughout the kitchen depending on personal preference.

♦ Kitchen Planning Mistakes

Do Not:

* Allow traffic patterns to cross the work triangle and interfere with the three primary work centers.
* Place the sink in a blind corner, too close to a wall reducing arm movement, in an island with little counter space, or too near the range.
* Install electrical outlets dangerously near water sources.
* Place the refrigerator next to the range.
* Install a built-in oven and range side by side. This is a fire hazard.
* Place appliances so that they block hot air and return air registers.
* Plan the isles narrower than 4 feet.
* Allow doors to swing into work areas or against appliances.
* Place shelves too high, or specify them too narrow for what they will hold.
* Select hard-to-care for, or non-heat-resistant counter surfaces.

NOTES

CHAPTER 19

• BATHROOMS

Planning your bathroom must start with an analysis of the family's needs in terms of the following considerations:

Family Size. The greater the number of family and friends using a bathroom, the larger it should be, or the greater the number of bathrooms required. This will mean additional floor space, linen storage, electrical outlets, and perhaps even more fixtures. When the same bathroom is to be used by family members of the opposite sex at the same time, compartment designs with doors can increase its utilization and reduce family conflicts. (see Figure 19-2).

Family Age. The ages of family members will also affect planning needs in a bathroom. Children may dictate the installation of hard-to-soil surfaces. The height of counters or lavatories may be governed by whether they will be used by children or adults. If the family includes elderly or disabled individuals, adaptions may include doors wider than 30", grab bars, higher bowl toilets, and other aids for their safety and convenience.

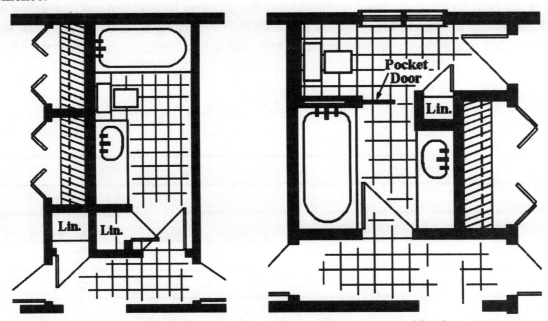

Figure 19-1. Typical bathroom layout.   **Figure 19-2.** Bathroom with privacy compartment.

Family Schedule. How many people leave for work or school at the same time? You may need multiple or compartmented baths. Two lavatories and a privacy toilet will allow a working couple to get ready for work at the same time in the morning. Within limits, plan facilities to be extensive enough, and arranged appropriately to ensure effective utilization of time and space.

Door Location. Check the doors swing direction to prevent inconvenience, or injury to those persons entering the bath, or using a fixture. Consider pocket doors for maximum utilization of bathroom space as they are designed to slide into wall openings, thus leaving more standing and working room. (see Figure 19-2).

Heating. No one likes to go from a warm bed or shower to a cold bathroom. Ensure that the heating duct is one inch larger in diameter if the bathroom has windows as this will keep the room cozy. Consider supplementary heat via heat lamps even if the bathroom does not have windows, as they speed up drying time, and keep you warm in the winter when reading your favorite book.

Ventilation. A bathroom generates more moisture than any other room in the house. Provide sufficient ventilation to prevent mildew, musty odors, peeling walls and ceilings, and fogging of mirrors. Specifically, an exhaust fan with sufficient capacity to handle eight air exchanges per hour is needed. The fan can be wired with the light switch, or independently. If the bathroom or powder room is small, and there is no window, wire the light and fan together. For the exhaust fan to work properly, provide intake air by allowing a 3/4" space from the finished floor to the under side of the door.

Electrical Outlets. Today's bathroom is a place that requires many conveniently located electrical outlets for hair dryers, hair curlers, timers for spa pumps and steam showers, shavers and all the other electrical gadgets that abound. Consider the users of the bath and their electrical needs, then install outlets and switches to meet these needs. Be sure the washroom wall outlets are ground faulted with reset buttons located in other washrooms to prevent electrical shocks - a very real possibility in bathrooms. Ground fault or reset buttons should also be located at the main electrical panel. All switches should be located so they cannot be reached from the tub or shower area. This not only insures safety, but is also a requirement of most building codes.

Mirrors. Mirrors, plenty of mirrors, can make a bath both beautiful and more functional. Big mirrors can make a small bath look larger; small or hinged mirrors properly located can aid in shaving and make-up applications. Mount mirrors so their tops are at least 72" above the floor so tall persons can see the tops of their heads. A full length door mounted mirror is very handy for checking pants or skirts.

NOTES

_____

_____

_____

_____

_____

_____

_____

_____

_____

_____

_____

_____

CHAPTER 20

• BEDROOMS

The three bedroom house is the most popular as it provides for a typical family of parents with children of both sexes. An individual spends more time in the bedroom than any other room in the house. Therefore, they should be designed to suite individual needs, and provide a separate space away from the rest of the family with the maximum amount of privacy and comfort for leisure and study periods.

Master Bedrooms usually require a zone design, since two adults although married require their own space, and as little interference from their mate as possible. Since furniture frequently occupies almost half the square footage of the bedroom with the remainder allocated to traffic flow, a 1 to 3 ratio of furniture area to floor area is recommended to allow for more movement space.

Bedrooms require sunshine and proper cross-ventilation. Consider window placement carefully to ensure they are not too large, too long and narrow, or too high off the floor to eliminate a view. Positioned properly, wall spaces between the window frames will accommodate the placement of furniture. Decisions on furniture placement in the room should be considered early in the design to determine the location of electrical, telephone and cable outlets.

The biggest complaint made by many couples is the lack of closet space provided when two adults share a room. With the use of closet organizers, closets can become less cluttered with each item having its own cubicle. Walk-in closets can also be used as private dressing areas when the wake up schedule differs. If this is the case, locate the master bathroom in close proximity to the closet and/or dressing area. A minimum rod and shelf space of 5 feet should be provided for each adult. A clear floor space of at least 54 inches for dressing should be provided between the closet and the closest piece of furniture. This will also allow for closet door swing without cramping the dressing area.

Design the master bedroom to include future growth of the family. If for convenience a baby is to share the parent's room, provisions should be made to accommodate a crib and a high boy dresser for clothes.

Secondary Bedrooms should be separated from the master bedroom by a sound barrier. Closets or bathrooms with insulated walls will achieve the desired effect.

Children very often wish to share bedrooms when young, as this often provides a safe and secure sleeping area, and keeps the boogie man away. When children share a room, bunk beds which can later be converted to single beds may be used to permit better use of space for playing and dressing. The conversion to single beds and bedrooms will be easier if one of the secondary rooms is designed on a larger scale so that the single beds can be arranged with sufficient, individual bed, play and closet space as well as privacy. Young children require storage facilities for clothes, toys, books, play table and writing supplies. Long term planning will encourage designing in space for a future students desk with proper lighting and outlet placement.

Provisions should also be made for laundry hampers, either in the bedrooms or a central collection area. When living in a two story home the designer may suggest a location for a laundry chute, so that clothes can then end up in the laundry room itself or in close proximity, perhaps a closet.

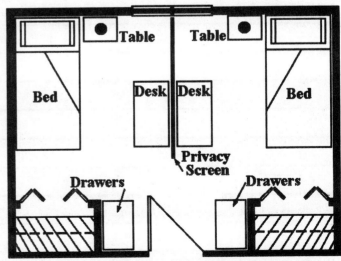

Figure 20-1. Room conversion from a bunk bed to two singles.

NOTES

CHAPTER 21

• LAUNDRY

Laundry facilities should be located in such a manner as to reduce the daily steps required to store and service dirty laundry. Check to ensure the layout includes sufficient temporary space in all bedrooms to store dirty linens and clothes, i.e., laundry hampers. If the house plan allows, a direct laundry chute will simplify the transfer of dirty clothes from the sleeping area to the laundry area.

If you have or are planning a large family consider the placement of the laundry facilities adjacent to the bedrooms, even on the second floor. This will greatly reduce constant trackage to an inconvenient laundry room. An additional closet space with increased depth is required to house the washer and dryer, and a floor drain with water tight flooring is mandatory. If space is tight, a stacked washer and dryer might be the answer in order to allow space for a soaking tub or small sorting counter.

Over the last 10 years lifestyles have changed and many home owners prefer a main floor laundry and mud room combination. Conveniently located for the working family, this room provides access to the garage, and is usually equipped with a washer, dryer, laundry tub or sink for soaking, counter space for sorting or folding, upper and lower cabinets for storage, and a closet to accommodate summer or winter clothes. With the return of fabrics requiring wet hanging, extra closet storage should be considered. It is always wise to have as many wall plugs as possible in this area, as they will come in handy. This room may also contain sufficient vertical storage for an ironing board or vacuum appliances.

The most convenient location for the laundry/mud room is adjacent to the kitchen, but this is not always complementary to the floor plan. An ironing center in or near the kitchen, and accessible to cooking or entertainment activities with the family is ideal. Convenient built-in boards may be considered, but the price sometimes does not justify the need.

NOTES

NOTES

_____

_____

_____

_____

_____

_____

_____

_____

_____

_____

_____

_____

_____

_____

_____

_____

_____

_____

_____

_____

_____

_____

_____

_____

_____

_____

CHAPTER 22

• STORAGE

Storage has always been a primary problem. There never seems to be sufficient storage space. The future enjoyment of a home will depend upon the time and money spent designing adequate and effective storage for items to be kept out of the way, in good condition, and yet readily accessible.

Day to day clothes such as jackets, hats, shoes and rain-wear should be kept in the mud room, back or front entrance. The front entrance closet should be large enough to contain visitor's coats and accessories.

Every day clothes and blankets which are kept in the bedroom closets change with each season. It is advisable to have a separate closet moth-proofed and cedar lined to protect these articles.

Linens should be stored adjacent to bedrooms and bathrooms with various depths of shelves for the different varieties of items.

Toys which are used most often should be stored in the child's bedroom. A bank of drawers or shelves can be contained in the closet, and they should be designed low and easy to open so the child can find and return it with minimal effort.

Hobbies and family games can be contained in a general purpose closet or cabinet in or near the family room for easy access. These items such as photo albums, scrap books, stamp or coin collections, puzzles, cameras, video tapes, cassettes or disks should be organized for daily or occasional use.

Cleaning equipment such as brooms, mops, dustpan, polishes and cleaners should be located centrally as they are used frequently in all parts of the house. Provide a storage space for tall items along with shelf space for the cleaning materials. Upstairs under-vanity storage for tub and toilet cleansers and cloths is usually sufficient for the bedroom area. Portable, rechargeable, wet/dry vacuum units require an adjacent wall plug wherever these units are to be located, i.e., in closets, under vanities or behind doors.

Sports equipment storage is again seasonal and requires easy accessibility and protection from the weather. They should be kept in a dry, clean place where they can be inspected and serviced when needed.

Overflow storage areas, are for possessions not currently in use that have been put away until needed at some future date. Space required for luggage, old trunks, and odds and ends of suspect sentimental value should be stored off the floor or on shelves in a dry, well ventilated room. This space will vary depending on the size of the family and the number of residing packrats. Speaking as a reformed packrat, it is best to get rid of these items at the dump or with a neighborhood garage sale before the move.

Garden tools and furniture are usually stored in the basement, the garage, or a portable garden shed which can be locked. A considerable amount of space is required for patio furniture, barbecue equipment, lawnmower, snowblower, weeder, garden hose, shovels or rakes and sometimes children's outdoor play toys.

Canned goods and preserves require a cool, and sometimes climate-controlled place so they can be kept without spoiling until used. For safe storage of bulk canned goods, preserves, wine, beer, and liquor items an insulated closet or room in the basement against an outside wall is sufficient.

Household tools are usually kept in a workshop, or an area of the garage that is well ventilated and heated. Depending on the handyman around the house, the space and storage needs will vary. Work benches, shelves, cabinets and peg boards are used for storing equipment and supplies. Extra care should be taken for the storage and ventilation of paints and solvents.

NOTES

SECTION 2: DESIGNING

CHAPTER 23

● CHOOSING YOUR DESIGNER

Note: *At this point start your diary logging every date, name, time and discussion.*

Word of mouth is usually the best way to find a designer. If friends were satisfied with the person who designed their home, chances are you will receive the same service. Ask the friend to allow you to review his set of blueprints, as this will provide a complete visual example of `working drawings´. The question you now have to ask, is "Will that same service be what you require to build your house?"

Another common method of finding a designer is spending some of your time on the phone. The local yellow pages can help you there. Usually found under Drafting Services or House Plans, these design services will vary in services provided, selection of plans for viewing, experience, quality of work and of course price. The larger more colorful ads will attract your attention, but do not ignore the small ads as large does not always mean the best in quality of design or price.

Many designers are using computers when designing houses. Computer Assisted Design and Drafting (CADD) provides the advantages of a sharper image, and faster modifications to the plans as they do not have to be erased and redrawn on paper. These services will be more expensive however, and the design remains only as good as the designer.

Since the standards and quality of the designer will determine the finished product, it is important for you to contact as many design services as possible, asking each designer the same questions.

◆ **Basic Questions:**
 * What are their hours - will you have to take time off work to see them?
 * How many years have they been in the business of designing homes?
 * Do they have a portfolio for you to review?
 * What do they provide as a full set of working drawings?
 * How long will the design take from start to finish?
 * How much will the complete design service cost with 8 sets of blueprints?
 * If there are any questions during construction will they be accessible?

You should then short list your selection to the best three, contact them again asking more detailed questions (see questions on page 68), and finally make an appointment to see them at their offices. This way you will have a first hand observation of their portfolio and people skills to know if you will be able to work closely together. Based on the answers and your gut feeling, make your final decision.

◆ **Detailed Questions:**
  * Will you be dealing with the owner or another designer?
  * Do they go out to inspect the lot prior to starting the presentation?
  * If during the presentation you are not satisfied with their work, and do not wish to continue, how much will it cost?
  * Does the designer live in a house designed/built by their design service?
  * How many contractors do they design for, and will these drawings be used by them in the future?
  * Can they provide you with a list of builders and subcontractors for quotations?
  * If required, will they do on-site inspections, and is this an additional cost?
  * Do they have any other services, eg., basement planning, landscape drawing?
  * Will the designs meet the Development Controls for the subdivision?

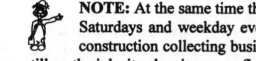

NOTE: At the same time that you are working with your designer it is a good idea to spend your Saturdays and weekday evenings (just after work) driving around the subdivisions still under construction collecting business cards of suppliers and subcontractors. Many of these people are still on the job site cleaning up or finishing the day's work.

If it is late in the day and there are no workers around, walk through some of the vacant homes under construction. You will find business cards nailed to wall studs or placed on window sills, or advertising signs on completed job sites. *Please keep in mind*, it is someone's house and private property. Respect their house as you would have them respect yours, i.e., leaving the house the way you found it, removing mud from your shoes before entering, closing the doors behind you, obeying no trespassing signs, and most importantly not smoking while in a house under construction!

◆ **INITIATING THE DESIGN PROCESS:**

Having chosen a designer that will suite your needs, make an appointment to discuss your rough draft material and start the presentations.

Multiple presentations are recommended if your information is in rough form, or not to scale as this provides the designer opportunity to suggest cost effective or structural changes. Presentations are also valuable for the client to take home for review and discussion with the family. For any required changes during this period, the designer should be updating his master drawings and providing you with blueprints of the changes. It may be necessary to have several meetings, to ensure all changes are fully understood and agreed upon as they could affect your over-all budget, and the final appearance of the home. Take each presentation home to review, and mark on it any changes that you feel are needed. If you do not understand something on the presentation, or have any questions about the construction process, make notes to be asked at the next meeting. Do not assume anything. The more knowledgeable you are at the presentation stage, the easier it will be to understand the final working drawings.The final presentation when completed should consist of blueprinted floor plans, and a front elevation for the streetscape visual. These presentation blueprints can then be used to obtain rough estimates of prices from suppliers or subcontractors. This will also tell you if the house is still within your price range. The suppliers or subcontractors can also review the presentation, and make any suggestions that might benefit or assist in keeping construction costs on line.

If at this stage there are only minor changes to the presentation, the designer can then proceed to the working drawings. However, if the suggested changes are substantial enough to change the structure of the house continue to request as many presentation drawings as necessary until you are totally satisfied with the floor plan(s) and streetscape visual.

Note: Review the checklist with the presentation drawings before proceeding with the working drawing.

♦ **Check List Before Working Drawings:**

In order to effectively evaluate the presentation, this list is designed to assist in asking questions of habitability and initiating family discussion. Putting all your trust in a Builder or Designer might misdirect you from your final goal. You and you alone should make the decisions for what you like or do not like, and what you want or do not want.

Exterior Streetscape:
- Do you find the house attractive?
- Are the walls, doors, and window areas visually well proportioned with the roof area?
- Will the roof shed snow, or are there valleys where ice can build-up causing potential structural damage?
- Has the designer shown flashing around the base of the chimney where it meets the roof, or where a wall and a roof meet? This prevents any unwanted water from entering. (see Figure 16-5 page 49).
- Have they made note of or registered sufficient roof ventilation for adequate air exchange in the attic area? Minimum requirements are 1 vent per 300 square feet.
- Are there more than three steps up to the front entrance? The fewer you see to climb the better.
- Is the front door entrance eye catching, and protected from the winter snow or summer sun and rain?
- Is the sidewalk and entrance well lit?
- Is the service door entrance visually exposed, or can it be protected with a gate and motion detector?
- Is the driveway to the garage straight and easy to maneuver without hitting a power box or light standard?
- Are there waterproof wall plugs by the garage and house for the use of electrical equipment or a car block heater?
- Can the driveway, sidewalk, and entrance be well lit with high visibility from the house to see who is approaching?
- Do the plans show hose taps (bibs) at the front of the house for watering the grass, and in or near the garage for washing the car to limit connecting multiple hose extensions?

Front Entrance:
- Is the front foyer large enough for three adults to stand comfortably with the front door and entry closet doors open?
- Does the foyer door open directly into a living environment, or is there a partition wall or railing separating the two?
- When the front door opens will there be a blast of cold air that would make you consider changing the door swing?
- Is the front entry closet in close proximity to the entry door?
- If the foyer is sunken is it possible for someone to trip and hurt themselves? If so, the foyer may be too small, or the step(s) too close to the door entrance.
- Is the foyer well lit with its own light and switch?

Mud/Laundry Entrance:
- Does it have sufficient counter space for shopping bags and parcels?
- Is the closet large enough to hold the bulky winter coats and provide space for boots?
- Is the floor material easy to maintain on a day to day basis?
- Does it have sufficient storage cabinets for your family's needs?
- Is the exterior landing well lit? Is there glass in the entry door? Is it safety glass?

- Is there a convenient area outside for the storage of garbage cans? Is it secured?

Living Area:
- Will the room fit your furniture, and allow sufficient circulation space for walking and cleaning without moving the furniture?
- Will there be an opening window close by for the circulation of fresh air?
- Does the window area provide sufficient privacy, yet allow the family members a good view of the entry walkway to see who is approaching?

Family Area:
- Is there sufficient wall space to fit your existing or planned furniture and entertainment needs?
- Does the fireplace location allow for furniture grouping with sufficient circulating space for the rest of the room?
- Does this room have a large window allowing the family to enjoy a view of a garden or deck area?
- Will there be good access to the front door, patio door and service door yet provide sufficient privacy from unwanted onlookers?
- Is the flooring you selected durable and easy to maintain?
- Will this area require a ceiling fan for air circulation or additional lighting?

Dining Area:
- Is the room large enough to fit your furniture, and allow sufficient room for expansion during family get-to-gethers?
- Can it be separated from other rooms of the house to keep its cleaning maintenance down?
- Does it provide sufficient privacy from passers by or neighboring windows?
- Will your existing furniture allow for good traffic circulation when people are seated?

Kitchen:
- Is it a workable kitchen when more than one person is moving about? Will they get in each others way?
- Is it well lit day and night, and easy to maintain?
- Is there sufficient storage for all appliances and groceries? Can they be easily organized for quick access?
- Is there sufficient counter space beside each major appliance as well as additional counter space to hold and operate small appliances, eg., coffee maker, juicers, and toasters, etc.?
- Do you have to purchase new major appliances, or will your existing units fit?
- Has the designer allowed for sufficient pantry and broom storage?
- Has the kitchen a conveniently located exhaust fan to vent cooking odors?
- Can the children's play area inside and outside be seen from the kitchen window?
- Does the kitchen allow quick, unobstructed access to the dining areas of the house?

Hallways:
- Are the hallways too long taking valuable room space from other parts of the house?
- Are the halls and stairways well lit with conveniently located 3 and 4 way switches? Imagine a night scenario when a family member having a restless sleep needs to access other areas of the house.
- Are the locations of the bedroom and bathroom doors in close proximity to each other, and away from hazardous stairs?

Bathrooms:
- Will the windows and fans provide sufficient air ventilation?
- Is there sufficient elbow space for shaving, brushing, combing, and hair blowing without feeling cramped for space?
- Is the size of the hot water tank sufficient when the shower is running, and someone turns on a tap or flushes a toilet in another part of the house? This test can be made when the plumber

has installed the fixtures. Has the designer registered the proper tank size? If you feel it is inadequate, have him indicate a larger size.
- Has the designer allowed sufficient - wall space for towel bars, medicine cabinets, mirrors;
                                       - storage space for towels, lotions;
                                       - floor space for laundry hampers, weigh scales, garbage baskets, etc.?
- Are the floors and walls easily maintained on a day to day basis?
- Is the room adequately lit and is there a sufficient number of ground fault plugs?

Bedrooms:
- Are the rooms large enough for your furniture, and will there be sufficient traffic space around the furniture?
- Will the windows provide sufficient daylight, evening privacy, and ventilation for the room without interfering with furniture placement?
- Has the designer provided sufficient, conveniently located electrical outlets for your intended furniture arrangement?
- Do the bedrooms have adequate closet space with a generously sized dressing area, and do the closet doors open wide enough for accessing the closet organizers without having unusable corners?

Basement:
- Will you be planning to develop the basement? If so, you might consider a 9 foot basement rather than the standard 8 foot.
- Are the locations of the structural posts going to limit the room sizes and locations? Consider moving them using structural beams.
- Are the furnaces, hot water tanks, floor drains, or sump pits located adjacent to or under the stairs where they will not affect future basement development?
- Does the basement have adequate ceiling lights and ventilation, and will the windows provide adequate natural light for the rooms when developed?
- Have provisions been made for a future 3 or 4 piece bathroom?
- Have you visualized where the storage areas are to be located for easy accessibility now, and to limit negative impact on future development.
- Has the designer shown proper floor slopes for drainage eliminating any puddling or basement flooding?

Stairs:
- Is their location central thus allowing good access from most areas of the house?
- Are they designed wide enough for moving large pieces of furniture?
- Has the designer allowed for the required headroom as per building standards?
- Are they too steep?
- Are the stairs safe for children? Can they be made child-safe by shutting them off with a door or gate?
- Will the vacuum hose reach the full length of the stairs (both up and down)?

General Questions For All Areas Of The House:
- Does the house require a laundry chute?
- Are the storage and closet spaces sufficient for the family needs?
- Will all electrical outlets be conveniently located for telephones, vacuum hose plugs, intercoms, security system, exterior motion censors, night light plugs, and those handy rechargeable flashlights and vacuums?
- Have you identified wall switch locations, and have you considered energy saving dimmer switches?
- Is the attic access in a convenient yet out of the way location, eg., in a closet or storage room, and not

a hallway? Is it easily accessible by a step ladder?
- Is the hot water tank size adequate for the needs of the family? A 2 bedroom house should have a 30 imperial gallon unit, a 3 bedroom house a 40 imperial gallon unit or two 30 imperial gallon units, a 4 bedroom house should have two 30 imperial gallon units.
- Are your exterior water, gas and electrical utility meters easily accessible for reading?
- Have the utility meter locations been discussed with your designer, and located on your drawing as per the Building Code, i.e., away from all opening windows and doors?
- Is the furnace thermostat(s) sheltered from drafts and sun which can drastically affect its function?
- Is the method of ventilating the rooms with fresh air satisfactory? Do the windows allow the intake of fresh air and the exhaust of stale air without causing blowing, drafty winds? Determine the wind directions, and have the window manufacturer locate the openers for proper venting.
- Are the heat and cold air return registers going to affect furniture placement? Can the designer indicate where they should be on the plan?
- Have fire prevention measures been considered? Should you have fire extinguishers? Are the smoke detectors properly located, and wired directly to the house power with a battery backup?
- Can you visualize how the family will escape if a fire occurs?
- Do bedroom and bathroom door locations allow for privacy from other main traffic areas?

Working Drawings:
It will be necessary to have several meetings at the working drawing stage, as this is where the structure and details of the home start to emerge. Request a meeting after the floor plans, sections, and elevations have been drawn, but before the final details such as dimensions, siding, window sizes, or required details have been completed. This step will allow you to review the final details of the plans without the confusing measurements, dimension lines and notes. This drawing stage will also allow you to plan for potential development of the basement by locating the structural posts and stairwells, furnaces, hot water tanks, sump pits and floor drains. Have the designer discuss with you how the basement can be best divided for future requirements.

Your designer should be visually walking you through the house plans explaining the structure, door swings, and room sizes with the proposed locations and types of switches, interior and exterior plugs, light fixtures, and any built-in systems required as a finished or rough-in product. At this time, if you remember some changes which have not been previously discussed at the presentation stage, and you wish to include them in the working drawings expect to pay for the additional work as they will require erasing the original plan. It is best to catch it at the drawing stage, and pay the little extra rather than during construction where it might not be possible, or could cost thousands to make the change.

Because of Building Code changes, or just working too quickly it is possible for even the best designer to make a mistake on the working drawings. By having several people review the presentations it is hoped that you, the designer, suppliers, or the subcontractors will catch any discrepancies in the house plans.

Assuming you have not missed anything, that all changes have been completed, and you are satisfied with the process, your designer can now complete the working drawings. At the last meeting, once again they will walk you through the house explaining all the final electrical, structural, general notes, specifications, finishing details, and revisions made to the drawings. If you require any additional notes or minor changes, speak now as this is your last chance for changes before blueprinting.

At the time of application for a building permit, the City Building Inspector will also have to review the working drawings before you can start construction. If the inspector finds something wrong with the working drawings, the liability of your designer is limited to making the necessary corrections with new sets of blueprints provided to you at no cost. If you have the house designed by an Architect, their design cost will usually include a fee for liability insurance. Discuss this potential problem and their policy with your designer before hiring them.

Your final meeting with your designer will most likely be the shortest. They will have the blueprint copies ready, and you will shake hands, and pay the first bill for your house, (see note below).

NOTE: In order to keep your finances in order, I suggest that all cheque payments on invoices come from one bank account, i.e., one especially set up for all payments made for house construction. All payments whether cash (via cheque made payable to yourself), or invoiced should be recorded by means of a cheque. The cheque should note what or who the payment was for, an invoice number, and whether it is a deposit, advance, or final payment made in full.

NOTES

_____

_____

_____

_____

_____

_____

_____

_____

_____

_____

_____

_____

_____

_____

_____

_____

_____

_____

_____

_____

_____

NOTES

SECTION 3: APPROVALS and ESTIMATES

CHAPTER 24

• STARTING THE PROCESS

◆ Surveyor:

The next step in the process is taking a set of blueprints to the surveyor for a plot plan. Request about 8 copies of the plot plan. Make sure you discuss with them the finished grades for drainage, the number of steps required at the garage and front entrance, floor heights from finished grade, basement height, driveway slope, depth and location of water and sewer lines, location and direction of the house on the lot, and finally the driveway length. When the plot plan is completed, usually in a few days, you will have all the necessary information to return to the bank to secure mortgage approval. At the same time take a set of blueprints to the city, pay the fee, and make application for construction.

Later in construction, after the house and garage foundation and framing are completed and backfilled, you will request the surveyor to return to verify that the house complies with the plot plan. When this is completed, they will provide you with a Real Property Report. Request at least 8 copies. This report states that the house complies with all city setback and sideyard requirements, and the house is within all property boundaries.

This Real Property Report is very important if you are planning to sell the house in the future. If this is forgotten and not done during the construction period, and the house for some reason does not comply with the surveyor's plot plan, you might have to pay a very large penalty to the city, or worse yet, have to move the house. A sale might be blocked because of this noncompliance so you might be living in the house for a long time whether you like it or not. The city engineering department will require a copy of this report for their records; the bank or the lawyer will require a copy for final mortgage registration on the title.

 Note: When this is all completed, request that the surveyor return your set of blueprints so they can be used to collect other price estimates. If they require a record of your plans have them take photo copies of the blueprints.

◆ Permits and Approvals:

By this time you may have chosen to build on your own, or selected a builder to build for you. Either way, the bank will require copies of the building contract, builder's specifications, and completed bank specifications which you have already completed along with blueprints, plot plans, construction insurance policy, and financial application as previously discussed in chapter 9.

If your subdivision has design controls, the control officer working for the developer will require one non-returnable set of blueprints for review. The bank and the city will each require 2 copies of the blueprints and plot plans, one of which is non-returnable. Once you have received their approval, they will each return one set of blueprints and plot plan with an approved building permit.

Before you receive the approved blueprints and plot plan from the city, the planning department will contact you by phone stating the plans are ready to be picked-up, and indicate that they have been approved, or approved with conditions. You will owe some money for the building permit, the fee usually being based on an estimated construction cost. Keep the approved plans returned by the city in a safe place. They should become part of the permanent record of the house along with mortgage documents, cancelled cheques, inspection reports, invoices, diary and any other documents pertaining to the house.

NOTES

CHAPTER 25

• SHOPPING FOR SUPPLIERS AND SUBCONTRACTORS

In order to get the best possible construction price for your home *you should get a minimum of three price estimates* from each of the suppliers and subcontractors required for the completion of your home. See the list below. Sometimes with the more expensive items (marked with an *) I suggest that even 4 or 5 estimates are advisable.

If you deviate from this system and decide that you are getting tired and two estimates are enough, or you are running out of blueprints this can be a BIG MISTAKE. For the little extra cost purchase more blueprints from your designer. If you do not, the final construction cost of your home could end up being thousands of dollars too high. I have made that mistake myself, and seen it happen dozens of times. It will be regretted, and can be avoided.

Not all of the suppliers or subcontractors require a full set of blueprints, so divide them up according to their areas of need. Start saving yourself about 100 dollars by dividing the blueprints, then save thousands with the suppliers and subcontractors by shopping smart.

How to divide your blueprints:

Sections = SE	Full Set of Blueprints = FS	Elevations = EL
Plot Plan = PP	Floor Plans = FP	Basement Plan = BP

Blueprint(s) required

Bank Mortgage	FS x 2 .... Will return 1 set
Surveyor	FS
City Permits	FS x 2 .... Will return 1 set
Excavator	PP, BP
Foundation Cribber	FP, PP, SE, BP
* Framer	FS
* Window Supplier	FP, EL
* Electrician	FP, EL, BP
* Plumber	FP, EL, BP
Light Fixture Supplier	FP, BP
Concrete Supplier	FP, SE, BP
Furnace/Heating Installer	FS
Drywaller/Taper/Insulator/Texture Installer	FS
* Painter	FS
Finisher	FP
* Lumber Supplier	FS
* Kitchen Cabinet Supplier	FP .... Photo Copies
Roofer	EL
Roof Truss Supplier	FS
* Siding/Stucco Installer	EL
* Carpet/Lino/Tile Installer	FP .... Photo Copies
Fascia/Soffit/ Eaves trough Installer	EL
Intercom/Security/Vacuum Installer	FP .... Photo Copies
Concrete driveway/Sidewalk/Floor Installer	PP, FP, BP

Fireplace Installer ....   FP .... Photo Copies

The above suppliers and subcontractors are the companies which account for about 92% of the cost of the house. The remainder such as appliances, weeping tile, water-proofing, railings, laminated beams and garbage removal, etc., will be discussed in more detail in chapters 26 and 27.

◆ COLLECTING AND RECEIVING ESTIMATES

This part of the construction process is probably the most time consuming and frustrating. Most of your time will be spent delivering blueprints for pricing. So that your time is well utilized prepare separate information packages in advance. Each package should have a label with the supplier's name, address and phone number stapled on their set of blueprints along with a list of your requirements. This list will ensure that they are all pricing similar products, and your final choice will be objective.

Remember, if you are subcontracting the house yourself you are considered a builder, and should be requesting builder prices from the suppliers and subcontractors. Which level of builder pricing will depend entirely on how much material you will be purchasing, and how many houses or subsequent referrals the suppliers and subcontractors will potentially receive from your business.

Some design services have negotiated special pricing with many of the subtrades you will be dealing with. As an example, many of my clients are able to contact a particular lumber yard, framer, window manufacturer, etc., and receive a class 1 or 2 discount, because over the years a sufficient number of my clients have requested estimates and purchased products from those companies. In return for these referrals I was given a contractor's rating, and my clients subsequently benefited from that purchase power. Ask if your designer has this service.

Calling and meeting with the suppliers and subcontractors is the second most important task other than the actual construction. This is where you can make or break your budget. Start by keeping your diary or day timer updated hourly with names, times and phone numbers. This will provide you with an accurate record of your discussion, the meeting time, date and place you met to discuss your information package. During this initial phone contact, try to get their home and or mobile phone numbers. This will insure that you are able to contact them if you run into any problems before any scheduled meeting or during construction..

The hours spent on the phone contracting suppliers and subcontractors is guaranteed to raise your blood pressure. If they are busy some (not all) will not show, will not call back, or show sometime in the next few days usually without calling first. This does not mean that you should strike them off your list. If they have a good reputation, but do not show after the second attempt, be persistent. Call them back. Ask them if they are interested in giving you an estimate as you do not want to waste their time or yours. You might find that continuous pressure will pay off. Human nature as you will find out comes with many variables that change on a day to day basis.

It is important that you inform the supplier or subcontractor that you are having other estimates done by other suppliers as well. It is only fair business practice so that they know, and can quote competitively. This will also ensure that they will give you their best price.

Before delivering the information package to the suppliers and subcontractors get a city map, and divide it into sectors. Mark on the map the location of each supplier and subcontractor, and then try to set up initial meetings in their order of proximity to each other. I have found that dividing the city into four sectors will speed and better organize your deliveries. Each sector represents one day to deliver the information packages; divide each of the sectors into two representing a morning and afternoon delivery. Make a habit of calling them a day in advance of each meeting, and set aside about one hour in the morning and at the end of each day to confirm or set new appointments for your next day.

Allow a fifth day for repeat deliveries due to the no shows. Try to book them in the morning, and if they do not show or have a good excuse the third time, and their reputation is not exceptional, forget them. They are not worth it! If you are unable to connect with them for the basic estimate, you can imagine the potential frustrations when it comes down to the actual work or installation. The afternoon of the fifth day will be used to deliver information packages (prepared in advance) for new meetings set up to replace those that did not show, or to get additional estimates.

During this initial meeting, request that they have the blueprints and estimate completed by the next week, and schedule that meeting at the same time as your present delivery if possible. Now that they have the full information package, and a date is set to pick-up the estimate, make sure that you remind them. Call them three days, then one day before the pick-up is scheduled. This reminder will ensure that it will be ready when you arrive, and also provides them with the opportunity to ask any questions that might have been missed during the first meeting.

The estimate should contain a detailed list similar to yours naming the product(s), materials and labor they are promising to supply. If they feel they have an item other suppliers do not provide, have them include it in the estimate for your consideration. When this happens, go back to the other subcontractors, and ask them to include the item in their estimate, or price it as an option. Make sure you know in advance who is responsible for supplying specific items before selecting your final estimate. Prices will vary drastically depending on the material you are expected to supply. Do not be fooled by cheap estimates that only give a single, non-detailed, total package price. In the end they might cost more, especially if you have to supply and pay for all the materials. This statement will be repeated several times in the following chapters where it was a significant issue when I was building. Also discuss the lien holdback, and their terms of payment requesting that it coincide with the mortgage draws. This should not come as a surprise to them, as it is common practice in the building industry.

You will find that many of the estimates you receive will have been calculated on the square footage. Double check that your designer has correctly calculated the *actual* living area square footage. Some designers consider dead areas such as chimney fireplace chases, or basement stairs to be part of the square footage. This should not be the case as they are not actual living and usable environments.

It is possible for each estimate to vary in price because of different name brands supplied, warranties, and after installation services. If you feel more comfortable with a particular company and product, but you prefer someone else's price, tell the preferred company of your predicament. Without giving the price of the lower bid ask if they can sharpen their pencil. You will often find that they are able to meet and sometimes beat the lower bid. If you are also lucky enough to be in a position to pay upon completion rather than the usual 30 days, they will sometimes offer you an additional discount. It never hurts to ask! Make sure you have a good warranty contract ,and you know the supplier will still be in business to service their work.

Note: Initially it will not be necessary to wait for all three estimates from *all* the suppliers or subtrades. During construction you will not be using some trades people, eg., painters, finishers, light fixtures, concrete drive-way/sidewalk installer, carpet/tile/lino or stucco and siding installers until later so that only one estimate will be sufficient to fix a price for construction costs for the mortgage application. During the initial construction phase you will have time to shop for the additional estimates from these suppliers and subcontractors, so do not panic.

When you have chosen the supplier or subcontractor, verify with a follow-up letter the price, service to be supplied, product to be supplied, payment schedule, holdbacks, service or product warranties, and notification of responsibility for the removal or clean up of any debris resulting from their service or crews. I will discuss the clean up responsibilities of each supplier or subcontractor in the following chapters.

For your convenience, on pages 81 to 86 I have provided sample quotation sheets. I have personally used this format to record the many estimates received for residential, multi-family and commercial construction projects. It is useful as a constant reminder for the number of estimates that have to be collected before proceeding with the construction.

NOTES

Job Number : _____ Address : _____ City : _____

LEGAL DESCRIPTION : Lot : _____ Block : _____ Plan : _____ - _____

Owner : _____ Phone No : (____) _____ - _____

BUILDING COSTS	DATE		COST ESTIMATE					FINAL NAME & PRICE		
100 - Plans and Design Fee										
101 - Building Permit										
102 - Surveyor Building Pocket										
103 - Surveyor Stake - out										
104 - Surveyor Grade Certificate										
ENGINEERING										
105 - Soil Test										
106 - Structural										
107 - Wood Foundation										
UTILITIES										
108 - Water Line & Trenching										
109 - Sewer Line & Trenching										
110 - Gas Line & Trenching										
111 - Temp. Electrical Service										
112 - Electrical Service										
113 - Utility Consumption										
114 - Gas Line Application										
115 - Winter Frost Allowance										
CONSTRUCTION COSTS										
116 - Excavation & Backfill										
117 - Dirt Fill or Disposal										
118 - Framing Lumber Package										
119 - Structural Beams										
120 - Roof Trusses										
121 - Prefabricated Wood Stairs										
122 - Deck Joist Lumber										
123 - Deck Railings										
124 - Deck Floor Material										
125 - Deck Stairs										
126 - Concrete Costs										
127 - Winter Costs										
128 - Concrete Pump										

COSTS CONTINUED	DATE		COST ESTIMATE					FINAL		
FOUNDATION LABOR										
129 - Footings										
130 - Foundation										
131 - Structural Pads										
132 - Grade Beam										
133 - Pilings										
134 - Foundation Reinf. Rebar										
CONCRETE FINISHER										
135 - Reinf. Rebar & Wire Mesh										
136 - Basement Floor										
137 - Garage Floor										
138 - Driveway Pad										
139 - Sidewalks										
140 - Patio										
141 - Steps										
142 - Precast Concrete Steps										
FILL SAND and SPREADING										
143 - Basement Sand										
144 - Garage Sand										
145 - Driveway & Sidewalk Sand										
146 - Foundation Waterproofing										
147 - Weeping Tile & Gravel										
148 - Sump Pit & Liner										
UPPER STRUCTURE										
149 - Rough Grading										
150 - Finished Grading										
151 - Spreading Black Dirt										
152 - House Framing Labor										
153 - Deck Framing Labor										
154 - Window & Door Caulking										
155 - Roofing Material										
156 - Roofing Labor										
157 - Window Cost										
158 - Exterior Doors										
159 - Storm Doors										
160 - Plumbing										

COSTS CONTINUED	DATE	COST ESTIMATE						FINAL		
161 - Heating										
162 - Air Conditioning										
163 - Electrical										
164 - Eaves Trough										
165 - Soffits - Fascia - Gutters										
166 - Batt Insulation										
167 - Loose Fill Attic Insulation										
168 - Rigid Insulation										
169 - Vapor Barrier										
170 - Drywall Material & Labor										
171 - Masonry Labor										
172 - Masonry Material										
173 - Steel Angle Iron										
174 - Siding and Installation										
175 - Stucco and Application										
176 - Parging and Application										
177 - Painting - Int. & Ext.										
178 - Finishing Labor										
179 - Finishing Materials										
180 - Cabinets										
181 - Carpets and Linoleum										
182 - Marble and Ceramic Tile										
183 - Wood Flooring										
184 - Wood Railings										
185 - Iron Railings										
186 - Wood Paneling										
187 - Mirrors										
188 - Medicine Cabinets										
189 - Bathroom Accessories										
190 - Light Fixtures										
191 - Ceiling Fans										
192 - Wall Papering										
193 - Garage Door										
194 - Garage Door Opener										
195 - Skylights										
196 - Shutters and Louvres, etc.										

COSTS CONTINUED	DATE	COST ESTIMATE							FINAL		
197 - Winter Propane Heating											
198 - Fireplace											
199 - Fireplace Mantle and Instal.											
200 - Security System											
201 - Intercom System											
202 - Exterior Security Lighting											
203 - Built - In Vacuum System											
204 - Sprinkler System											
205 - Garage Unit Heater											
MISCELLANEOUS											
206 - Supervision											
207 - Travel and Fuel											
208 - Additional Labor											
209 - Cellular Phone											
210 - Phone Bills											
211 - Developer Damage Deposit											
212 - Rental Equipment											
213 - Garbage Removal											
214 - House Cleaning											
215 - Furnace and Duct Cleaning											
216 - Window Cleaning											
217 - House and Constr. Insurance											
218 - Workers Compensation											
219 - Cable TV Connection											
220 - Telephone Connection											
221 - Delivery Costs											
222 - House Inspection Service											
LEGAL and MORTGAGES											
223 - Legal Fees											
224 - Bank Interest											
225 - Property Taxes											
226 - Offsite Levies											
227 - Land Cost											
228 - Land Interest											
229 - Mortgage application fee											
230 - Bank Appraisal Fee											

COSTS CONTINUED	DATE		COST ESTIMATE						FINAL		
231 - Mortgage Insurance Fee											
232 - Mortgage Draw Interest											
233 - Interim Financing Costs											
234 - Selling Costs											
APPLIANCES											
235 - Range Hood											
236 - Dishwasher											
237 - Range											
238 - Refrigerator											
239 - Washer & Dryer											
240 - Barbecue											
241 - Cook Top Stove											
242 - Micro - Wave Oven											
243 - Trash Compactor											
244 - Freezer											
EXTRAS											
245 - Drapery											
246 - Venetians											
247 - Fence											
248 - Patio											
249 - Retaining Walls											
250 - Sod, Trees & Shrubs											
251 - Washed Rock, Cedar Chips											
252 - Septic System											
253 - Water Well											
254 - Water Softener											
255 - Sauna and Heater											
256 - Steam Shower Heater											
257 - Hot Tub or Spa											
258 - House Address Numbers											
259 - Moving Costs											
260 - Hotel Costs											
261 - Furnace Air Filters											
262 - Special Humidifiers											
263 - Heat Exchanger											
264 - Water Hoses											

COSTS CONTINUED	DATE		COST ESTIMATE						FINAL		
265 - Lawn Mower											
266 - Snow Blower											
267 - Power Attic Fans											
268 - Garage & House Door Weather Stripping											
269 - Federal Taxes											
270 - 5% Contingency Fee											
DEDUCTIONS											
271 - Federal Tax Rebates											
272 -											
273 -											
274 -											
275 -											
ADDITIONAL COSTS											
276 -											
277 -											
278 -											
279 -											
280 -											
281 -											
282 -											
283 -											
284 -											
285 -											
286 -											
287 -											
288 -											
289 -											
290 -											
291 -											
292 -											
293 -											
294 -											
295 -											
296 -											
T O T A L S :											

SECTION 4: PROJECT MANAGEMENT

CHAPTER 26

• WHAT TO EXPECT FROM SUPPLIERS AND SUBCONTRACTORS

The next chapter will be a description of what each supplier and subtrade should be expected to provide as a service, what options can be considered, and what you as a consumer should look for when requesting estimates, purchasing materials and clarifying responsibilities during the construction period.

A smart shopper should know as much as possible about a service or product before purchasing. If you have any questions or doubts about a product ask the salesperson for a reference address so you can see the product as it would look on an existing house.

◆ DESIGNER

The designer can provide you with many services other than the presentation and working blueprints for your home. Many design services deal with suppliers and builders on a daily basis. This direct knowledge can be a benefit if they provide you with a list of reputable subtrades willing to give you estimates for your house.

In addition, for a small fee, they can coordinate on-site inspections with you, and advise you of any visual problems to be corrected, or energy saving features that you might include during the construction. With a good set of house plans and a knowledgeable designer the construction of your home can hopefully be an enjoyable and a less troublesome experience.

◆ SURVEYOR

Surveyors are usually computerized for convenience and accuracy of drawings. Once you have chosen your lot, ask them to provide you with a building pocket of your lot. This building pocket will tell you the width, size, shape, and the most suitable placement of the house on the lot. A building pocket will greatly help your designer, and eliminate any possibility of zoning and subdivision infractions.

Ask your surveyor for assistance if you wish to raise your house, i.e., out of the ground to eliminate any window wells, or plan any special landscaping for the future. They will position and adjust floor heights to best suit your needs in conjunction with sewer heights, grading, and lot draining, and consideration of your neighbor's positioning.

When requesting your survey, have it marked and staked the morning of the excavation. That will eliminate vandals moving your survey pegs resulting in a re-survey, or worst yet an incorrect excavation.

 Note: The first house on the street usually determines the neighboring floor heights and house placements on the lots for proper streetscape.

◆ MORTGAGE INSPECTOR

Mortgage inspectors control the amount of monies advanced at completion of certain construction phases. These inspections are done when an inspection request card has been completed and mailed, or by phoning the inspection department or mortgage officer to request an on-site inspection. Be sure that you have complied with all the construction phases noted on the card before you request the inspection. If you have not completed the phases you will not get your advance, and you will be charged for an additional inspection when you have complied. See chapter 15 page 45 for samples of the inspection request form.

These inspectors differ from the city inspectors whereby they approve the completion of specific construction stages and certain building requirements, eg., frost insulation under garage foundations and basement frost walls. Building code compliances are left to the city building inspectors.

◆ CITY INSPECTOR

City inspectors can be either your best asset or your worst enemy. Before starting the construction review your documents from the city to understand their requirements, and know when to request inspections and schedule meetings with the inspector placed in charge of reviewing your house plans. They can inform you if any of your neighbors ran into unforeseen problems such as ground water or poor soil conditions which will then require you to provide them with an engineer's report. Knowing these potential problems will save you time and money.

Having these people on your side can be your best asset. They are there to inspect the workmanship, and make sure your house plans, heating, plumbing, electrical, insulation, house structure, grading, etc., have been done correctly, and according to the local building code. They will review your house plans and plot plan making sure that the materials are correct and structural requirements are satisfied.

When you pay for your building permit at the city hall these inspection services are included. If you have any questions or concerns during construction, request an inspection. They would be more than pleased to help, and will inform you of any deficiencies that can be more easily corrected at that time rather than later at great expense.

◆ EXCAVATION CONTRACTOR

Excavation contractors use the heavy equipment such as a caterpillar, back hoe or grade all to dig the foundation hole. With the plot plan in hand and survey stakes located properly on your lot, they will excavate to the foundation depth predetermined by the surveyor. To achieve this the excavator uses a surveyor's transit and a grade marker to confirm the final foundation depth. This depth must also be to undisturbed soil.

Note: If there is an existing city sidewalk and curb have the excavator protect them by covering them with some soil. The dirt should be spread wide enough to accommodate the width of a dump truck, and thick enough to withstand the weight of the truck when fully loaded. This seemingly unimportant request will limit damage to the curb and sidewalk, and avert the city service crew repairing the damage and sending you an unexpected and expensive bill.

Excavation might sound very simple, but it requires a man who operates a machine that removes about four cubic yards of dirt at a time being able to excavate a level hole to within 1/2 to 1 inch of its required depth. As they remove the earth from the foundation's hole they directly load the unwanted earth into waiting trucks which transport the dirt to a dump site. The dump sight should be located as close as possible to keep the truck transportation costs down. It might be a good idea to ask some of your neighbors, the developer, or local builders if they know of any local dump sights. The excavator must also determine how much earth should remain around the perimeter of the excavation providing sufficient material to backfill around the foundation for proper grade slope when required. It is also essential that they leave several areas open with sufficient space for foundation and concrete trucks to gain access to the perimeter of the hole. That does take some experience and skill.

When the weeping tile has been completely installed, inspected, and approved, the excavator will return to backfill the hole using a small caterpillar. He must backfill the hole on an angle to the house so the weeping tile will not be crushed. Also, if there are any clay lumps they will be pushed in on an angle to the foundation rather than directly at the foundation. If not backfilled correctly the clay could cause the still fresh and curing concrete foundation to crack or cave in. The house is slowly and carefully backfilled first at each corner, and then, still on the angle, backfilled to the middle of the foundation walls.

The foundation must be totally backfilled with a proper slope to drain away from the house. With the settling of the perimeter earth over the next 10 to 12 months, there will most likely be uneven settling. The excavator should leave sufficient hard packed clay or dirt for himself or another operator to return, and fill and tamp the areas that require additional fill dirt. This left over fill will be used at a later date to guarantee proper drainage of the lot as well as the required sloping grade away from the house foundation for effective water runoff.

Some excavation contractors will at the same time be using a back hoe to excavate and install your sewer and water lines. Ask if this is possible as it will save you some time by eliminating an extra subcontractor's estimate. Make sure the subcontractor knows where the garage and sidewalk supporting piles will be located, so that he can keep the water and sewer lines away from these areas. This will prevent the lines from being crushed or hit by the pile auger resulting in the lines having to be dug up and repaired.

◆ ENGINEER

Engineers are sometimes required to do soil tests before the footings can be formed. The size of the footings are usually determined by the designer as per requirements of the local building codes. However, the city inspector sometimes requires you to contact an engineer should the condition of the soil in neighboring excavations contain too much water, or be too unstable for the footing size specified on the blueprints. In these cases the engineer will have to be on the construction site near the end of the excavation to collect soil samples. When this has already happened with other excavations in the area, the engineer will already know the soil conditions, and be able to determine the required size of footing to be used for that specific soil type.

The city inspector will usually accept the engineer's footing size being given verbally on-site, but will require a written soil test and report from the engineer for their files. This way construction can proceed without any delay. Be prepared in advance by asking the city inspector, neighbors or the developer who sold you the lot.

◆ FOUNDATION CONTRACTOR

Foundation contractors or cribbers like to be present prior to completion of the excavation in order to double check the foundation depth. If the foundation depth is too deep they will require the excavator to re-fill the hole to the proper depth, and tamp the clay to an undisturbed condition. This fortunately does not happen too often. When it does, the excavator bears the expense unless the surveyor is in error, or someone had moved the surveyor's stakes.

With the plot plan and a surveyor's transit the cribber determines the height and location of the foundation footings and structural pads. Most cribbers supply their own nails and footing forms because the forms can be reused several times. It is the responsibility of the owner to supply a box of coated nails and pegs with which the forms are nailed. Some times bracing materials are required to hold some sections of the forms together. If the cribber has to cut his forms for this, you will be required to replace them. Make sure you know in advance who is supplying what before selecting your final estimate as prices will vary drastically depending on how much material you supply. Do not be fooled by cheap estimates that only give a final price, as in the end they might cost more especially if you have to supply and pay for all the materials.

The day the footings are poured make sure you tell the cribber to insert a proper keyway into the footings. A keyway will eliminate any side movement of the foundation during and after backfill. (see Figure 26-1). After the footing forms are removed, the cribbers will install the vertical foundation forms. They will be held together with metal snap ties which they supply. The owner/builder will be supplying the horizontal reinforcing rods, window bucks, and perimeter 2"x 4" nailers to be placed in the foundation forms prior to pouring. Prefabricated basement window bucks can be supplied by your window manufacturer, but they have a reputation for poor construction. Usually put together with staples they occasionally fail when pouring the concrete. A better idea is to supply the material from your lumber yard, and have your cribber make them. He will charge you about the same price as the window manufacturer, but they will be made to his specifications, and he is ultimately responsible if they fail during the foundation pour.

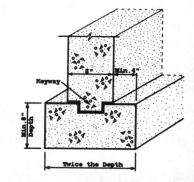

Figure 26-1. Footing sizes

To pour the concrete for the footings the cribber uses wheelbarrows, shovels and wooden chutes. If the chutes of the concrete truck are not long enough, a concrete pump has to be used to pour the foundation. The length of the pump's boom is determined by the distance required to pour the concrete into the farthest part of the foundation. Concrete pumps have a very expensive hourly charge, therefore ask the cribber to order the pump so that he can schedule it with the concrete supplier.

Note: Ask your cribber if he requires a concrete pump when you receive his estimate so that you can make the necessary phone calls to get several estimates for a pump.

Basement walls can be poured without the floor joists, therefore it is good practice to brace walls over 30 feet in length to eliminate movement. Many cribbers include the framing and placing of the structural basement beams, floor joists, structural walls, structural wood, and steel posts in their service. This helps to support the foundation, and eliminates movement when pouring the concrete. Most cribbers are not to be considered framers, but the more years of experience they have the better they become at framing floor joists. The crew is only as good as the owner/supervisor. If you find that the owner/supervisor has been on his own less than three years, ask him to split his estimate into one price for the foundation, and a second with floor joists included. This way you have the option of having the cribber or framer placing your floor joists. Regardless of who you choose to set your joists, it is the cribber's responsibility to make sure that the foundation and grade beam is square and level. The leveling is usually done with the foundation forms in place just prior to pouring the walls.

When the concrete foundation and grade beam walls are being poured, a worker must vibrate the concrete to eliminate the chance of air pockets in the foundation around the reinforcing or the window bucks. Air pockets reduce the strength of the concrete's structural capability, and if this happens, the gaps or "coning" will have to be patched and repaired by the cribber using a special hand mix concrete.

Schedule the cribber to return after the backfilling has been completed. At this time he will stake the piles stipulated on the drawings to be augured for the garage floor, garage grade beam, sidewalk and deck. Several holes will be drilled in the house foundation to accept reinforcing rods where the garage grade beam and the house foundation connect. Once the pile holes have been augured, they will set up the forms for the grade beam, and place the required reinforcing rods in the piles, grade beam, and pre-drilled holes in the house foundation. The concrete delivery should be scheduled directly after the forms have been installed and levelled, and the reinforcing completed. This way the grade beam and all the piles can be poured at the same time. (see Figures 26-5 and 26-6).

The location of the piles are carefully specified by the designer. The piles are positioned in areas underneath the garage floor where the automobiles weight can be best transferred onto the supporting piles and grade beam rather than the garage floor. This will reduce the potential cracking of the garage floor caused by the vehicles.

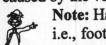

Note: Have the cribber do a q̲... i.e., footings, foundation, grade...

Stairs can be included in the s̲... supplier's quantities, and if they are way o̲... adjusted before you start building.

◆ LUMBER SUPPLIER

Lumber suppliers tend to treat the customer as if t̲... feeling when you first meet them, but you soon realize... tight competition with other lumber suppliers. They requ̲... that they can order new. In order to give you the best prices... bulk, and this requires customers like yourself ordering con̲... material turnover. You will find that every estimate from each ̲... quantity and price. Some estimates have more wall material or r̲... ...sts or plywood. In addition many lumber yards charge for estimates.

An accurate estimate is the best estimate, i.e., where every stick and ... a 5% waste factor added on. This, however, is usually not calculated into the estim̲... are actually best <u>guestimates</u>. If they do a proper estimate they would lose customers. First, b̲... estimates would be so exact that the total cost of the package would be higher than their compet̲...s, and second the client could take the estimate to a competitor, and without lifting a finger the competitor would guarantee the same price. The first business then has only a fifty-fifty chance of getting the contract, but has invested 100% of the time and effort. Therefore, the lumber yards estimate low so that the price is low, and they have a better chance of getting the contract.

The lumber and finishing packages in total are the most expensive items in a house, and this is where shopping smart will benefit the purchaser. With blueprints in hand, go to the selected lumber yards, and ask for the best <u>contractor's price</u> possible for lumber, finishing materials and finishing hardware. Then ask for their monthly price list. These price lists are calculated on a price per thousand board measure or per item. When the time comes, regardless of the lumber yard chosen, that is how you will be invoiced. Go to each lumber supplier and ask for the same price list. Review all the price lists, and compare the price per item or per thousand board measure for each individual material type, i.e., 2" x 4" and 6" studs, 2" x 10" and 12" floor joists, 4' x 8' by 1/2" and 3/8" sheets of plywood, etc., making sure you are comparing the same grades of materials, i.e., fir, spruce or pine.

Establish who has the lower all around prices for most of the lumber and finishing materials, especially the expensive items, and that is the company you want to buy from. Go back, and tell them you will be purchasing your full house package from them based on their price list, and you would like to open an account. Ask for a material estimate for your house package, (you will most likely get it for free if you will be purchasing from them), and how long the price will be honored. After reviewing the estimate and comparing it to the other supplier's price lists, ask if they will give you a better price on the more expensive items (no harm in asking), and honor that price at the time you are building. Have all that written on the estimate.

Note: When selecting lumber suppliers, make sure they carry the full line of construction materials so you will not have to run around to purchase unstocked items at other stores. This will waste your time, and most likely require direct cash payments if you do not have an account with the other supplier. Also, lumber yards usually have a delivery fee per load, so try to keep the delivery numbers down by following the construction schedule in chapter 31.

...he dirtiest job during the construction period. It is their job to ...ater is directed away from the foundation through 6" diameter plastic ...ary sewer line or a sump pit. What they connect to depends on the local ...ther connection the basement walls and floor should remain dry.

...e requires that they place weeping tile around the outside perimeter of the footings, and ...gravel around the tile to eliminate for a period of years sand and silt from blocking the water ...side the tile. (see Figure 26-2).

When collecting estimates for the weeping tile have them include installation of the sump pit and window well drains to collect the ground water in the quote. Discuss the location of the sump pit with the designer, so that one of the future rooms in the developed basement will not have a sump access hole in the floor. Alternate locations are in a basement laundry area, furnace room, or better yet underneath the stairs as these areas will not have carpeting. The weeping tile suppliers do not connect the footing pipe to the sump pit as that is the job of the plumber, and usually done when they install the basement drains and water lines. The sump pit can also be installed by the plumber, however, they usually prefer someone else do it.

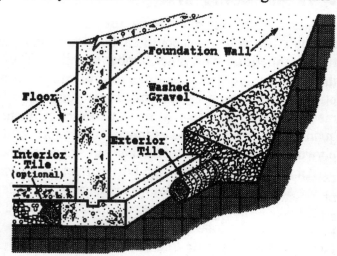

Figure 26-2. Illustration showing perimeter weeping tiles around exterior and interior of footing.

Note: If you are planning to put the sump in or near the center of the basement, it is a good idea to have the plumber install a check valve in the sump's drain line. This will stop the backflow of water and the gurgling noise that goes along with it when the water in the line backflows into the collection pit.

Window well drains remove the collection of water around the foundation wall at the frost level thus eliminating the potential problem of ice and frost shifting and cracking the foundation walls. (see Figure 26-3). These vertical pipes are placed at the bottom of the window well, attached to the foundation, and connected to the weeping tiles around the footing. Washed rock is placed over the drain tile at the bottom of the well.

Since they are hired to eliminate water problems they also apply the water-proofing to the foundation walls. This water-proofing is a brushed and sprayed coat of tar to the face of the foundation, i.e., brushed where the cribbers snap ties remain to eliminate seepage, and sprayed to approximately 1 foot above where any backfilled earth will be in contact with the foundation wall. Some building codes and mortgage companies require that this waterproofing also be applied to the interior walls of the foundation. Check

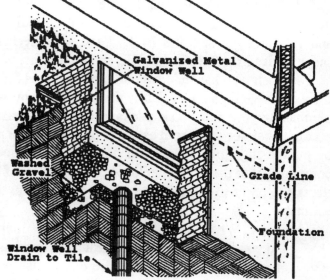

Figure 26-3. Vertical window well weeping tile to footing.

to see if this has to be done in your area before you collect your estimates.

Check with your city building inspector to see if there has been any ground water problems, or if a sump pit is required for your area. Additional precautions should be taken if the ground water is high. You will know this problem exists when the cribber pours his footings, and the excavation fills with water, pools, or the base becomes very muddy and too sloppy to work in. One way of eliminating this major problem is by having additional weeping tile placed around the inside perimeter of the footings, and having that line directed toward the sewer or sump. It is an extra expense, but it will eliminate future water damage and very costly repairs. (see Figure 3-2).

♦ **CONCRETE SUPPLIER**

Concrete suppliers differ very slightly in their selections of concrete available, and are all able to supply the different concrete strengths and custom aggregates required. The difference would be in the aggregate size and color which is dependant on the gravel yard supplying their raw materials.

Prices will vary within 2 - 8 dollars per cubic meter or yard of concrete which does not make a significant difference when building one house. The important difference is the delivery times available as they prefer to service the larger contractors first unless booked a day in advance. Some suppliers will not deliver before 7 am, after 5 pm, or on Saturdays. When collecting estimates find out which supplier will provide the best service and the best price. Have the supplier quote each concrete aggregate with the price and strengths available for the different aggregate types. Also provide the supplier with the required set of blueprints for a quantity estimate, and have him separate the estimates into the eight categories, i.e., footings and pads, foundation, grade beam, piles, garage pad, basement floor, driveway, and sidewalk, (include the stair material in the sidewalk quantity).

There are five common types of concrete. Starting from Portland type 10 they increase by 10's to the highest sulphur resistant mix, type 50. Each mix is used during different seasons and for different weather conditions with heat, calcium, and air added selectively to each mix to improve strength and setting times. Calcium and heat are added to accelerate the setting time of concrete during cold weather. The amounts to be added are determined by the concrete supplier, and is dependant on its use and the current temperature. During extremely cold weather, i.e., around the freezing point, scaffolds with tarpaulins and propane powered heaters will be required to cover and heat the concrete during the pouring and curing process. These extra costs will have to be considered, and discussed with your supplier when pouring in cold weather.

Air is added to concrete where foundations and floors will be in contact with water. It creates a medium that acts as a sponge allowing the mix to contain more water, and then allows it to evaporate without destroying the structural integrity of the concrete. A garage floor, even though protected by a polyethylene vapor barrier, reinforcing rods, and a roof, still requires air as it is exposed to water dripping from cars, and the melting snow. The floor surface has a slight slope, and has to be power troweled to a smooth and shiny surface which allows for proper water drainage without pooling. Water collection will cause frost damage during the cold months of winter.

Basement floors do not require air as the vapor barrier will not allow water to penetrate. The basement is also heated, and therefore any water that might come in contact with it will quickly evaporate. The surface requires a slight slope to a basement drain, and has to be power troweled for proper water drainage without pooling in the event of a water pipe ever bursting.

For house construction you might require three concrete strengths, and their selection will depend upon the amount of reinforcing, present weather conditions, and structural requirements of the area to be poured. In general, 2000 psi. (15.0 mpa), 2500 psi. (17.5 mpa), or 3000 psi. (20.7 mpa) are used for the following areas of house construction:

Garage pad .... 3000 psi. (20.7 mpa)

Footings, pads, piles, driveway and walkway .... 2500 psi. (17.5 mpa)

Foundation, grade beam and basement floor .... 2000 psi. (15.0 mpa)

See appendix chart for metric/imperial conversions on pages 193 and 194 if required.

If you have already chosen your cribber, check for the name of his preferred supplier. Ask if he has an account with them, or has any connections when you open an account. If you have already opened an account, but the cribber likes to deal with a different supplier, phone that company to see if they will honor or better your supplier's price. If they can, open an account with them, because they are all so similar it does not matter who you purchase from, as long as your cribber is satisfied with the one you have chosen.

Because the pouring of concrete is so dependent on the weather, have the cribber organize the ordering and deliveries to your site. Most cribbers like to pour concrete early in the morning, in fact the earlier the better. If your cribber has ordered concrete for 7:30 am., and at 6:00 it is raining very hard, chances are he will not be pouring concrete that morning. He will then phone the order desk of the supplier, cancel the order, and request another time in the day just in case the weather changes. If the weather remains poor he can always cancel the order again. Concrete suppliers are used to this happening, and do not mind as long as they have 90 minutes or more notice.

♦ CONCRETE FINISHER

Concrete finishers pour and place flat concrete surfaces, i.e., the basement floors, garage floors, driveways, sidewalks and custom specialty items such as exposed aggregate steps. They perform an extremely valuable service. They are a special breed, requiring the patience of Job, and it seems the ability to discuss all the qualities of concrete, the weather and politics for many hours without stopping. When it comes time to pour or power trowel the concrete you had better get out of their way, as they work like nobody else I have ever seen. Really good finishers who care about their work are worth having on your team.

Concrete finishers do not supply the sand, reinforcing rods, wire mesh, vapor barrier, pegs or forms. It is their job to provide only the labor for cutting the reinforcing and wire mesh, spreading and tamping the sand, cutting and placing the poly vapor barriers, drilling the holes in the foundations for the dowels which tie the floor to the foundation, installing the stair, walkway and driveway forms, and finally, placing and finishing the concrete. The only items supplied, other than the required nails and operating equipment, are the chemical retarder and sealer which are used when placing and finishing exposed aggregate finishes. The charge is on a cost per square foot basis for placing and finishing the concrete, and a flat rate for spreading and tamping the sand.

 Note: Before ordering sand for the basement, make sure that all the rough-in plumbing, floor drains, sump pit, sewer and water lines, and under-slab heating are completed and inspected by the mortgage and city inspectors, and all the subcontractors have removed their debris from the basement (as included in their estimates).

In the next section I will discuss some of the more important points to remember when working with concrete finishing.

Basement floor:

After the sand has been delivered (through the basement windows), the finisher spreads it evenly so that when compacted it will be the same thickness as the footings and structural pads. A string and level are used to make sure the sand is uniform after being compacted with a power vibrator. A 6 mil. polyethylene vapor barrier is laid over the sand, structural pads, and footings to eliminate water penetration into the concrete. Narrow sheets of plywood are laid on top of the poly to protect it from tears when using the wheelbarrow to spread the concrete. A surveyor's level and measure are used to make sure the concrete floor is a minimum 3 inches thick at the drain, and allows for a proper slope to the drain. The finisher will be using a very long handled, large trowel called a bull float which forces the water in the concrete to the surface to create a smooth and even surface. Once this process has been completed, the finisher must wait until the concrete has set sufficiently before using a power trowel. The concrete is trowelled until hard, smooth and shiny. (see Figure 26-4). The concrete sometimes is required to contain 1 or 2% calcium to speed setting.

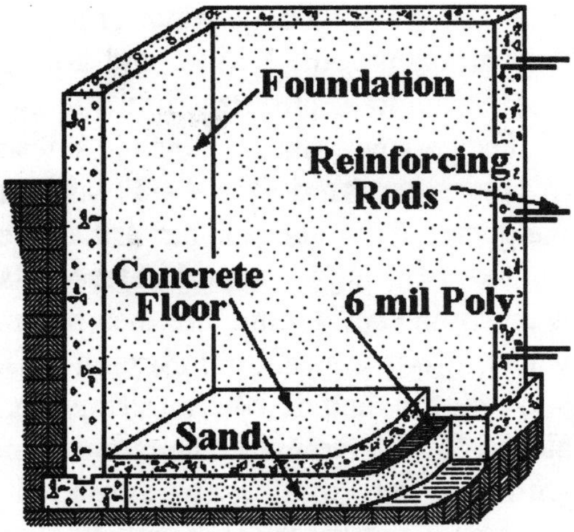

Figure 26-4. Basement floor drawing.

Garage floor:

Using a chalk line the finisher marks the top of the concrete floor, which is determined from the level of all the rough garage door accesses, and ensures a proper slope to drain. Measuring for a 3 - 4" concrete floor, the finisher drills holes approximately 1 1/2" below the chalk line and 24" apart all around the interior perimeter of the garage grade beam. Steel dowels approximately 24" long are then hammered into these perimeter holes to which the floors 6"x 6" 10/10 wire mesh reinforcing will be connected later by wire ties. (see Figure 26-5).

After the sand has been delivered and spread, a string and level are used to ensure the sand is uniform, and allows for the slope to drain before being compacted to garage floor thickness with a power vibrator. If possible, the sand is sprinkled with water to achieve its maximum compression.

Then a 6 mil. polyethylene vapor barrier is laid over the sand to eliminate water penetration into the concrete floor. Wire mesh is now laid on top of the poly, taking care that it does not puncture the vapor barrier, and connected to the perimeter reinforcing dowels which tie the floor to the foundation. As the concrete is being spread over the poly the finisher will pull up the wire mesh and the rebar from the piles to make sure they will be positioned in the middle of the concrete floor to increase its structural stability. As with the basement floor the finisher will be using a bull float to force the water to the surface producing a smooth and even appearance. Once again the finisher will wait until the concrete has sufficiently set before using the power trowel. The concrete is trowelled until hard, smooth, and shiny. Depending on the wind, weather, and finisher the concrete sometimes contains 1% calcium to speed setting, and is ordered with air to

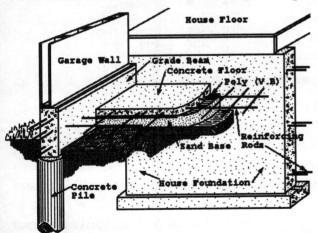

Figure 26-5. Garage floor connected to grade beam.

speed up the evaporation of water.

Driveway and Walkways:

The finisher again uses a chalk line to determine the top of the driveway and walkway. The walkway is measured from the front garage floor height, and allows a proper slope to drain from the garage floor to the bottom of the house entry steps; the driveway is measured from the front garage floor height to the top of the city curb or sidewalk with a slope predetermined by the surveyor's plot plan.

Walkways that are attached to the garage and house should be supported every 8 feet on the outside perimeter by concrete piles to eliminate sinking. In areas where the soil has been disturbed, or drilling piles might break or damage water or sewer lines, an alternate method is to pour supporting concrete brackets. The brackets are 8" wide, and are placed directly over and connected to every grade beam pile with reinforcing to ensure structural strength. (see Figure 26-6).

The concrete finisher sets the sidewalk and driveway support forms high enough to allow for at least 5" of compacted sand for a 4 - 5" thick concrete walkway, and a 4 - 6" thick concrete driveway. Similar to the garage floor, the finisher drills holes approximately 2" down from the chalk line and 24" apart around the exterior, adjacent perimeter walls of the garage grade beam and house foundation. Steel dowels, approximately 24" long are then hammered into the perimeter holes to be connected to the driveway and walkway concrete and reinforcing rods.

After the sand delivery, a string and level are used to make sure the sand has been uniformly spread, and allows for the required concrete thickness. The sand, as for the basement and garage floors, is compacted with a power vibrator, and sprinkled with water for maximum compression. A polyethylene vapor barrier is not required for any exterior concrete as water should be allowed to penetrate the concrete and flow through the sand to undisturbed soil. This will carry any water away from the foundation.

Steel reinforcing rods or rebar 20 feet in length and 3/8" to 1/2" in diameter are laid on top of the sand, and connected to the perimeter dowels with wire ties which structurally secure the driveway and walkway to the grade beam and foundation. Steel dowels 24" in length must always be used to attach the driveway, garage pad and walkway to the grade beam, however, wire mesh can be used to replace the 20 foot reinforcing rods. Adding wire mesh to the rebar in areas where the driveway and sidewalk connect will structurally strengthen these usually weak areas, and reduce substantially the future chance of cracking.

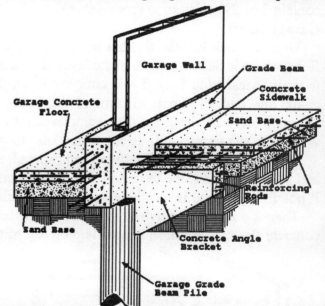

Figure 26-6. Sidewalk being supported by a angle bracket.

As the concrete is being spread over the sand the reinforcing rods are pulled up to make sure they will be contained in the middle of the concrete thus increasing its structural stability. The finisher, as with all floors, uses a bull float to force the water in the concrete to the surface, and with short back and forth motions smooths and levels the concrete. Depending on current weather conditions the finisher will patiently wait for hours until the concrete has sufficiently set before roughening the surface with a broom to produce a nonskid surface.

The finisher's last task is to place a finished edge around the perimeter of the walkway and driveway, and the expansion joints placed every 10 feet. Additional expansion joints on the walkway are located where there might be potential crackage.

Exposed Aggregate:

This finish requires special care and accurate timing by the concrete finisher to achieve the desired result. The concrete supplier has a special size and color of gravel that is used to produce the rough and colorful exposed rock surface.

The same process for a standard driveway or walkway is followed for pouring this concrete. However, once finished with the bull float, a retarder solution is evenly sprayed over the concrete which stops the top layer of concrete from curing, and allows the course, colored gravel to stand out. They cannot spray too much or too little, as too much retarder will permeate too deeply stopping the curing of the lower layers of the concrete, and too little will not allow enough of the top layer of concrete to stop curing. Once sprayed the concrete must be watched closely for up to 3 to 8 hours, or until the retarder is ready to be removed, and the concrete base has cured enough to be walked on. Clouds, sun or even shade from an overhang will vary the curing time of the concrete, and the time required for the retarder to work. When the finisher sees that the concrete and retarder are ready, the retarder must be washed off as quickly as possible. A water tap and hose must be available for the finisher to wash down the concrete, and then a broom to sweep the retarder and the uncured concrete skim.

Because the exposed aggregate is too large, expansion joints are not placed by the finisher. Instead, expansion joints must be saw cut into the concrete (see the next section on concrete cutting). The exposed aggregate is also considered the final product so the finisher does not use a trowel to provide a finished edge as he would with a brushed concrete finish.

After the expansion joints have been completed, and the concrete sufficiently cured in strength, the finisher will return to give the surface a diluted acid wash. The brushed-on acid cleans the rocks and removes any excess concrete powder still remaining. The remaining acid is then brushed and hosed off to expose a clean and porous surface. This surface must be protected from the elements by sealing the surface with an epoxy or plastic-like finish which is brushed or rolled on. Once completed it brings out the natural colors of the exposed aggregate, and protects the concrete surface with an almost shiny, acrylic polish.

Exposed concrete requires additional work, therefore expect the concrete finisher to charge more for the job. If your budget can afford it it is well worth it, and gives the streetscape of the house a real touch of class.

The above finishing processes are similar at each construction stage, but change with the variables used such as the concrete aggregate type, air and water content, and reinforcing and final sealing requirements. For all jobs include in the estimate that the finisher is responsible for stripping the forms once the concrete has completely set. You can give the forms to the finisher, have him purchase them for a cheap price, or have him throw them on your material trash pile.

◆ CONCRETE CUTTING

Exposed aggregate requires expansion joints every 10 feet on the driveway surface, and every 5 feet on the walkway where potential cracking might occur. A high speed, self moving concrete saw is required for cutting the expansion joints in the driveway. Closely following the chalk lines marked on the concrete the concrete finisher uses a high speed, water-cooled cutting blade to penetrate about 1 1/2 inches into the concrete. The worker directing this machine takes his time to make sure the line being cut into the concrete is straight. This machine is only able to cut expansion joints on large surfaces such as driveways, as it is too long and bulky to cut sidewalk expansion joints.

For sidewalks, tight corners and limited access, a hand held concrete cutter is used. The hand cutter does not have its own water supply so the operator waters down the concrete as much as possible to control the dust. This does not eliminate the dust blowing all over the concrete surface which is why the concrete finisher does the acid wash after the concrete saw procedure is finished.

Concrete cutters charge by the lineal foot for all expansion joints, so for budget planning count on about 200 lineal feet of expansion joints for a regular driveway and sidewalk.

◆ SAND/GRAVEL SUPPLIER

Provide them with the square footage and average depth of sand required for each area, and they will calculate the number of square meters or yards of sand required. Because they charge per meter/yard of sand delivered with a surcharge for partial loads, have the basement and garage sand delivered at one time.

The basement and garage work areas are protected by the roof structures. Water is the worst enemy of sand. If not protected from rain, the sand could wash away, or become too wet making it impossible to spread and level by hand. Several days might be wasted before the sand dries and is spreadable. Also double check the amounts to be delivered with the supplier and the finisher, as it is better having extra sand than not enough.

Timing the delivery of the sand for your driveways and walkways is very crucial. Have the concrete finisher order the sand for you as he will make sure the weather conditions are favorable, and the sand will arrive ready for spreading when they are at the job site. Suppliers who specialize in the delivery of sand for basements and garages use a conveyor belt attached to the end of their truck to toss the sand through the windows and doors. Request them to place sand in the hard to reach areas to minimize the amount the finisher will have to use the wheel-barrow. He and his helpers will appreciate it.

◆ FRAMER

Word of mouth is the best way to find a good framer, but the final choice should be made knowing **1)** years of experience, **2)** price, **3)** number of crew, and **4)** conscientiousness to clients needs. Once you find a good framer, be honest about your budget, and he will try to be flexible within reason with his price.

The more work experience someone has the better the quality of work, and also the less time it will take to finish. The downside is the more experienced the framer the more it will cost, however, do not let the price difference scare you too much. Most framers know what their competition is charging, especially if working for general contractors who will pay only so much to a framer. (If a final price has been contracted with a client, the contractor will expect the framer to be more flexible in his pricing.} Competition for work is usually in the builder's favor as a framer who wishes to keep his crew must keep them busy. Even during boom economies there seems to be a good selection as framing crews tend to migrate around the country to find the jobs.

Framers charge by the square footage of the house with extra charges for fireplace chase, bay windows, decks and balconies, or other items that require on-site framing. Before framing the house it is important to review the estimate, and make sure the following has been included, especially "that he will return as needed to move or add any floor joists, bracing, backing and wall studs for the electrician, drywaller, finisher, plumber or heating contractor". Discuss with the framer the type, location and installation of vapor barriers, and the insulation requirements for corners and hard to get areas that might be inaccessible to the insulator once framing has been completed.

As discussed earlier, either the cribber or the framer can place the floor joists, but before choosing make sure the cribber has the experience. It makes for a cleaner job if

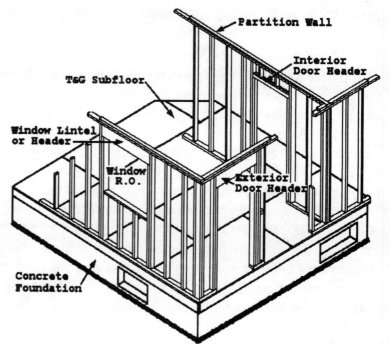

Figure 26-7. Illustration showing typical framing structures.

you have the framer place the joists, because he knows exactly where the structural walls will go so the joists will be more suitably placed. (see Figure 26-7).

Assume that the framer was chosen to layout the joists, and the cribber has already installed the structural beams, posts, bridging, and joists in their places. (see Figure 26-8). The framer after reviewing the floor plans will start selecting the lumber to be used for the joisting. Request him to use only the joists that are not twisted, split or damaged when delivered, and to provide a lumber count of the material to be returned to the lumber yard for credit and exchange for better material. Give the lumber yard the framer's lumber count so they can deliver the new floor joists when they pick up the unusable lumber. It is a good idea to have your framer inspect all deliveries, and reject any lumber or materials that are not up to quality standard.

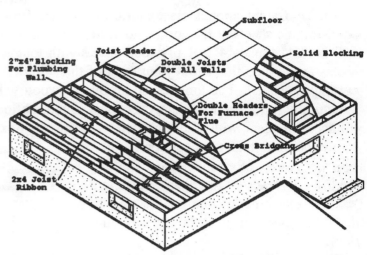

Figure 26-8. Drawing showing floor joist framing with subfloor.

On completion of the floor joist layout, and prior to cross bridging and subfloor installation, do a quick review with the framer. Make sure that there will be sufficient headroom created at the stairs (see Figure 26-11), and sufficient insulation placed in all foundation floor joist voids that will not be accessible after the subfloor is in place, and vapor barrier and insulation installed. Once these items have been visually inspected and approved, the top supports of the cross bridging will be permanently nailed leaving the bottom supports free to be completed (nailed) once the house has been completely framed with roof sheathing, windows, and doors installed. Most cantilevers are framed using only standard floor joists around the perimeter header. This will not allow sufficient insulation in the joist area to keep the cantilever properly heated and warm during winter months. (see Figure 26-9). Bay, bow and box windows are usually

cantilevered as it is much cheaper to extend the floor joists beyond the house foundation for these areas than build a very costly concrete footing and foundation. The distance the joists can project beyond the foundation may vary in every state or province across North America, and depends on the building and snow loads it can structurally support. Your designer will know these cantilever requirements. (see Figure. 26-10).

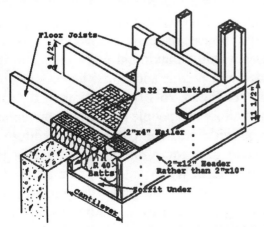

Figure 26-9. Illustration showing how typical cantilevers are insulated and framed.

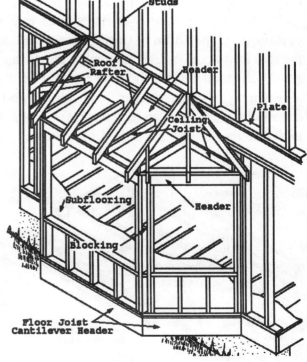

Figure 26-10. Framing of a typical cantilevered Bay window.

As the framer applies elastomeric glue to the floor joists, his crew will be laying the subfloor and screwing the sheets at the corners to secure them in place. The most common subfloor material comes in 5/8" or 3/4" thick - 4' x 8' plywood sheets with tongue and groove ends for a continuous, tight fit. The sheets are laid so the 4 foot joint ends of the subfloor are directly over a joist to minimize waste when cutting the sheets. The tongue and groove plywood will stabilize the floor joists making the floor more structurally rigid for the walls. After a few of the subfloor sheets have been placed, the framer will anchor the sheets in place with additional screws before the flexibility of the glue is lost. Request him to place the screws no more than 6 inches apart.

Note: Remind the framing contractor to place sufficient elastomeric glue on the tops of the floor joists prior to laying any subfloor.

Today, the parts for the main stairways are usually fabricated in a stair shop or millwork plants, and then installed at the job site by the framer.

The main stairways leading to the upper floors are usually installed after the framer has completed framing the house. This eliminates any damage to the stairs that might result during the framing and rough-in plumbing, heating and electrical stages. Basement stairs should be installed

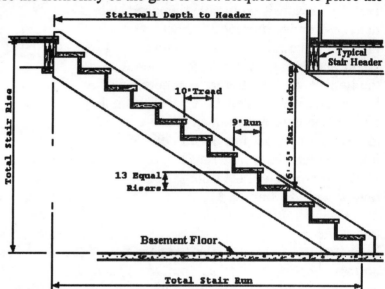

Figure 26-11. Illustration showing typical stair construction.

after the basement floor has been poured and has had time to cure. The framers will build temporary stairs from framing lumber to provide easy access to the different levels of the house.

Stairs consist primarily of risers and treads supported by stringers. The height of the riser is called the unit rise, and the width of tread (nosing included) is called the unit run. The sum of all the risers (usually divided into equal measurements) is the distance from a given floor to the next floor up or down, and the sum of the total tread is the total run.(see Figure. 26-11).

Framing of stairs requires that trimmers and headers in the rough framing of the floor joists be doubled especially when the span is greater than 4 feet. Headers more than 6 feet in length should be installed with joist hangers unless supported by a beam, post, or partition. (see Figure. 26-12). Providing adequate headroom often constitutes a problem, especially in smaller house structures. Installing an auxiliary header within a maximum of 24 inches to the main header will permit additional headroom above the stair. (see Figure 26-13). The width of the main stair should allow two people to pass without contact, and also provide sufficient space so furniture can be moved up or down. A minimum of 3 feet is generally recommended. When walking up or down a stairwell a person must have the opportunity for support. This is usually provided by a continuous handrail along one side.

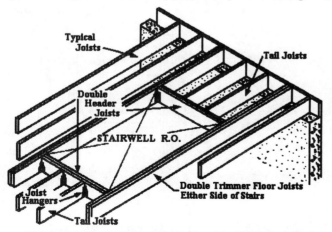

Figure 26-12. Drawing showing typical stair framing.

A complete set of working drawings should in-

include detailed drawings, in section, of the stair system especially if the stair layout has a landing at any level. All the stairs will also be shown on the floor plans showing the correct number of risers and treads.

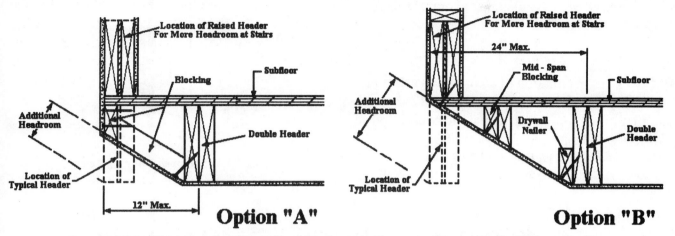

Figure 26-13. Illustration showing optional framing procedures to gain more headroom at stairwells.

Points to consider on a walk-through:
- Is the handrail well attached to the wall?
- Are the light switches located correctly, and easy to reach?
- Are there light switches at the top and bottom of the stairwell, and is the light properly located for lumination of the stairs?
- If the railing has spindles are they close enough together so a child's head will not get caught?
- Will the vacuum hose reach the full length of the stairs, both up and down?
- Has your framer double checked the headroom to see that it conforms to code?
- Check that the stairs are securely attached to the walls to eliminate squeaking.
- Should you consider a keyed lock for the basement door as security or child safety?
- If you have a pet, should you provide an access hatch so they can enter and exit the basement as required?

It is important that the upper floor framing be completed and roofed as quickly as possible to eliminate warping. Once the roof, windows, and doors are completed, the rain and snow will have little effect on the lumber. There will be some lumber that will have to be replaced or braced as some twisting will occur as the lumber dries.

The wall framing of a house is standard throughout the industry, and the inspector makes sure everything complies with the building code. Here are some things to look for when walking through your house:

1) Use a carpenter's level to make sure all walls are level and vertical especially where any two corners meet. (see Figure 26-14).

2) Check to ensure sufficient insulation has been placed where interior walls meet exterior walls, and two exterior walls meet at corners. (see Figure 26-16).

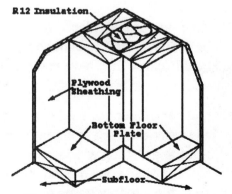

Figure 26-14. Illustration showing where two corners meet.

3) Check that the framer has installed a drywall nailer strip where walls meet at more than 90°. (see Figure 26-15).

4) Check for installation of a continuous strip of 6 mill polyethylene vapor barrier and a drywall nailer on all top plates, and at the connection of interior and exterior parallel walls. (see Figures 26-16 and 26-17).

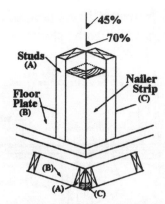

Figure 26-15. Illustration showing a greater than a 90° corner with nailer strip.

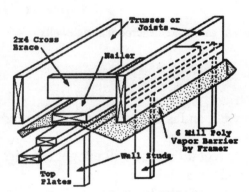

Figure 26-16. Vapor barrier and nailer at top wall plates.

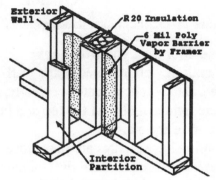

Figure 26-17. Vapor barrier at interior and exterior wall connections.

5) Discuss with the framer the rough opening of the windows and doors, and make sure that no more than 3/4 of an inch is added to the total width and height of each window when the rough openings are framed. (see Figure 26-18). Have the framer include in his estimate that if the windows are on back order, or if they have been broken and have to be returned, that he will return to install them. In the mean-time he will cover the openings with a polyethylene vapor barrier.

   Windows are usually placed on back order if they are out of the ordinary, or if the window manufacturers are backlogged because they have taken on too many contracts.

6) Check that the framer has stapled a strip of tar paper around the exterior rough opening of the window. Before actual installation of the windows and doors, have him put a bead of flexible silicone around the perimeter of the tar paper where the window frame will be nailed to the exte-rior wall. This will eliminate any wind drafts, especially if

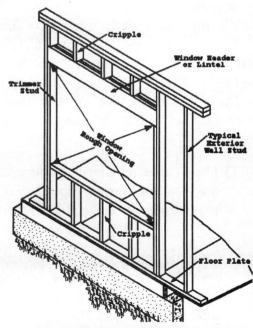

Figure 26-18. Illustration showing typical window rough opening.

the intended exterior finish is stucco. (see Figure 26-19).

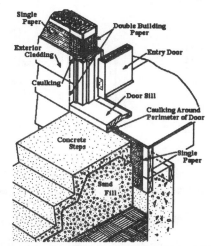

Figure 26-19. Illustrations showing caulking around perimeter of door and window frames.

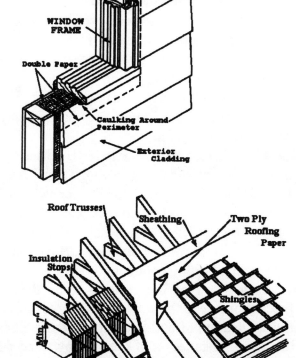

Figure 26-20. Baffle location drawing.

7) Do a visual walk around the house, making sure that all the attic insulation stops and baffles have been installed where required. If not completed correctly, there will be potential heat loss or water damage caused by the wind blowing the attic insulation away, and snow melting at the roof perimeter. (see Figure 26-20).

Unless agreed otherwise in writing, it is the responsibility of the framer to provide the necessary equipment and/or manpower to place the trusses on the structure for proper roof installation. Confirm with the framer the type of roofing material to be installed, and ensure installation of the proper type of outlooker for the roof structure. Adequate roof outlooker strength must be in place to carry the roof load set by the Local building code. The truss supplier will supply all the necessary roof trusses expect for those areas that require stick framing, i.e., fascia boards, outlookers, saddles, braces, and furred/ dropped down ceilings, etc.

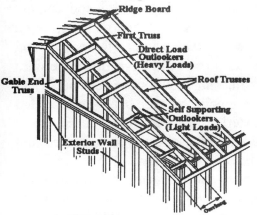

Figure 26-21. Drawing showing locations of outlookers.

 Note: In order to utilize as much of the on-site materials as possible, the framer will use the wall bracing lumber to construct the outlookers. (see Figure 26-21).

After completing the roof sheathing have the framer level the house by adjusting the basement steel teleposts, and bolting the posts to the structural beams with 2" lag bolts. Then he will complete nailing the cross bridging in the basement and second floor if you have one.

Review the locations of any framing and backing requirements for any wall fixtures such as towel bars, grab bars, toilet paper holders, soap dishes, medicine cabinets, wall and attic access hatches, spa pump service hatch, and other heavy wall and ceiling light fixtures or fans that might be required now or in the future. Have all these jobs completed before they leave your job site, because there will be little or no chance of getting them back before the house is ready for insulating and drywalling. If these items were not included in the estimate, it will be very hard to get

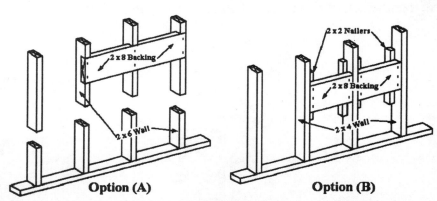

Figure 26-22. Illustration showing backing locations: **(A)** - 2x6 wall. **(B)** - 2x4 wall.

them to return especially if they have been paid in full, and you may have to do them yourself, or hire someone else to do it.

At the same time, check to see if he has made provisions at the attic and roof for the attachment of sheet metal tie straps and metal firestops which are required by code for furnace and fireplace flues. The framing for heat lamp boxes, snow and water saddles for chimney chases, and the roof cut outs for plumbing stacks or attic vents should also be visually checked and approved by the city inspector.

 Note: The framer when on-site cuts any holes for dryer and cook top vents that go through the outside wall or floor joist. Have him cut the hole as close as possible to the duct size indicated by the heating contractor.

As a result of the earth's ozone depletion and subsequent weather pattern changes, additional precautions need to be taken during construction for wind, snow and water protection. After completing the application of building paper around all exterior rough openings for windows and doors, and prior to the actual installation of the windows and doors, supply the framer with a quantity of flexible caulking compound. This will be applied on top of the building paper around the window and door openings, as well as on all exterior vertical inside wall corners. (see Figure 26-19 and page 132 for more details). This is an additional cost, but a very important item which will eliminate any water and air leaking into the house. This caulking application is usually not done by the framer, but if you supply the materials it would be simple for him to do the caulking while installing the windows and doors so have him include this application in his estimate.

To illustrate the importance of this item, I was constructing a multi-angled house, and got so involved with the completion schedule that I forgot to have the windows, doors, and inside plywood corners caulked. The house was completed, and no problems encountered until a very rainy and windy spring a year later. It rained continuously with high driving winds for several weeks which forced the rain up and underneath the flashing over some of the windows and doors, and caused the water to travel down the sides of the frames, under the door sills, and onto the floors. Normally the flashing would have been sufficient to stop the rain from getting in, but the high driving winds were just too much. Once the bad weather ended and the house dried up, the finisher was contacted to remove all the exterior door and window casings and add the caulking that should have been applied during construction. The casings were replaced and repainted. The remainder of that year was off-and-on wind and rain, but no water leaks.

At the beginning of the construction establish with the framer and other subcontractors an area set aside for lumber scraps and garbage collection. This way the framing crew will throw any lumber scraps on one pile, and not scatter them all around the house. It should be the responsibility of the framer to remove any debris, and sweep the house floor when he has completed his contract.

Construction hint: Many people when building their house cannot concurrently afford to install their deck, and usually plan to have it done sometime in the future when more affordable. Consider however preparing your house for the deck's future installation by having the framer install the 2"x10" wall header to the house where the deck will be attached. This way the exterior cladding installers will flash the top of the header, and install the cladding to the header. You will then have the header water-proofed and installed in preparation for the deck to be attached later. If not done by the framer during construction, the cladding installers will apply the exterior finish to the underside of the exterior sheathing. When ready for the deck to be installed, the exterior finish will have to be removed, the header installed with the flashing strip, and the exterior finish replaced to the top of the deck header. If stucco was removed it could become very costly and difficult to match the existing color. This pre-planning will eliminate unnecessary future expenses and duplication of work.

Do not play the game: There are very few areas where the subcontractor's jobs overlap. If they do, you will have to make a decision, i.e., which subcontractor is responsible to complete or repair what, and then stick with your decision once made. A good rule to follow when this problem occurs is anything having to do with plumbing the plumber will do, anything having to do with electrical the electrician will do, and the same goes for the heating, painter, finisher, etc. The same stand should be taken when damage is done, i.e., cross bridging, studs, or backing being removed and not replaced. Whoever broke or removed the item should replace it, or put in some other structural material to support or to take its place. If they do not honor your decision, tell their boss, and write a note informing him that you will determine who is responsible, have someone else fix it, and charge it back to the guilty party. Send them a photocopy of the repair invoice when it is taken off their final invoice.

◆ ROOF TRUSS SUPPLIER

Most of the roof structure is constructed with pre-manufactured roof trusses designed to the specifications shown on the house plan blueprints. The simpler the roof line the cheaper the overall truss costs, therefore try to design the house with as many standardized roof trusses as possible. The more trusses that are the same, the fewer times the manufacturer has to change his truss template, and hence the cheaper cost. When more roof lines are added to the inside or outside, or the roof slope increased above a normal 4/12 slope, the more difficult the trusses are to build. There will also be extra costs incurred from having to have each different truss engineered. (see Figure 26-23).

As there are different roof types, as discussed in Chapter 16, there are different roof trusses available to create the different roof lines. (see Figure 26-25). Any one truss style may have many design and structural variations. The structural capacity, design and cost will vary with different truss spans, roof slopes, roof snow load, truss spacing, end heel height, grade of material it is manufactured from, and weight bearing function, eg., carrying other truss loads. There are also many names used for different parts of a truss such as heel height, gussets, cords, etc.

Trusses in the past were designed to be supported by a wall or structure at each end, however, given the

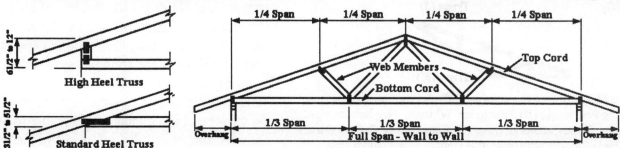

Figure 26-23. Illustration of a typical Fink or "W" truss with optional high and standard heel heights.

more complicated roof lines and current house styles, they have been adapted to allow cantilevers. The design of the cantilevered length of truss will be determined by the local ground snow loads, size of the top and bottom cords, roof slope, and bracing location or size. (see Figure 26-24).

When collecting the estimates, ensure the package price includes on-site delivery, repair of any damaged trusses up to the delivery date, all connecting devices, and a list of all the individual trusses detailing their span lengths, slopes and number delivered. The manufacturer must provide engineered drawings for each individual truss with the engineer's seal, or a covering letter stating that all the trusses have been engineered. These authorized drawings or covering letter will be required by the city engineering department for their files before you get your final building permit. This paper work is required even though you have already received approval to start the construction.

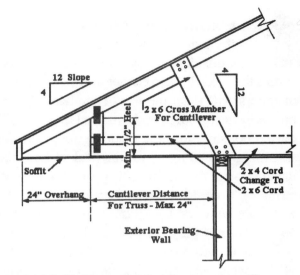

Figure 26-24. Illustration showing a truss cantilever.

To illustrate the importance of damaged product, a careless delivery of trusses unfortunately caused one of my clients a week delay in the framing of their house. It had been raining for a few days prior to the delivery date, and the construction site was drying up, but still muddy. The trusses were to be delivered at 8:30 in the morning, so my client arranged for his father-in-law to wait at the job site to show them where the framer wanted them placed, i.e., at the front of the house where the ground was relatively dry. The father-in-law waited until 10:00, then decided to drive over to a local convenience store to phone the truss supplier. The supplier apologized, and said that the delivery had been delayed for an hour or so, but the driver should have been there and gone. When the father-in-law arrived back at the site the driver of the delivery truck had somehow gotten to the back of the lot, and was proceeding to unload in the muddiest area. Unloading consisted of reversing the truck, and quickly braking so the trusses would slide off the flat bed of the truck. The trusses did slide off the truck, but rather than sliding off gently, they dropped and broke apart sliding into the foundation of the house resulting in even more breakage.

Luckily it was witnessed, and the driver of the truck, and the owner of the company were verbally and colorfully informed of damages and responsibilities. The next day the owner

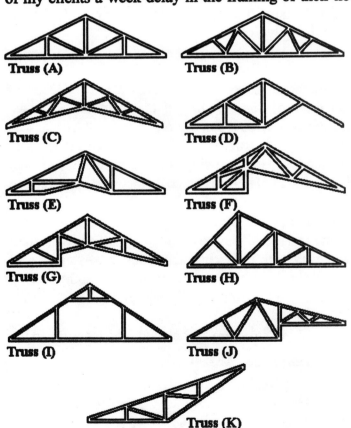

Figure 26-25. Illustration showing different roof truss types.
(A) King Post truss. **(B)** Triangular 'W' truss. **(C)** Scissor truss.
(D) Outer Slope Cathedral truss. **(E)** Center Slope Cathedral truss.
(F) Single Slope Studio truss. **(G)** Double Slope Studio truss.
(H) Dual Pitch truss. **(I)** Attic truss. **(J)** Step truss. **(K)** Inverted truss.

sent one of his sales representatives out to inspect the damage, and take a damaged truss count. Four days later the new trusses were delivered, and the damaged trusses removed. This unfortunate problem, however, cost my client 5 days delay, i.e., five days of interest payments to the bank. He contacted me and asked me what he should do. I told him to record it in his diary, have his father-in-law immediately write down exactly what he saw, send a letter of intent for payment to the truss supplier with an invoice for property and structural damages, deduct that amount from his estimate, and try to settle out of court. The end result was rather favorable, because the owner of the truss company agreed to pay for all damages, and had the property re-graded once the ground had dried up. The only irretrievable thing was the 5 days lost in construction time, but all construction will incur a few weeks of delays caused by this type of damage or poor weather.

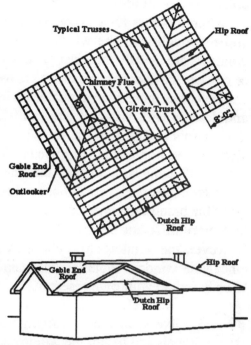

Note: As you see a cellular phone can be an invaluable item on a job site.

The truss manufacturer must also supply the framer with a detailed truss drawing showing the layout and location of each individual truss, and their connection to each other. This way there will not be any mistakes with extra trusses left over or trusses not delivered. (see Figure 26-26). Framers get exasperated with truss manufacturers that do not include the location of <u>all</u> valley, gable end, hip jack, and extension trusses in their layout. Your house designer will usually register on his drawing only stick framed areas that cannot be supplied by the truss manufacturer. Your framer will give you an estimate including

Figure 26-26. Upper - Truss drawing showing different roof types for elevation - below.

all stick framing and gable end "lookout" soffit framing based on the working drawings, and the truss manufacturer should do the same.

For clients to better understand the many truss applications for any one house, ask to see a truss chart that shows the different possible types of trusses. It becomes much easier to understand what the room or roof slope will be with a visual presentation. The truss drawings, (see Figure 26-25), and the assistance of your designer will help you to better understand and select the most appropriate truss styles.

♦ **WINDOW SUPPLIER**

Windows are a necessary and important part of your home. They are necessary to get light and air into your home, but they are also an important part of the architectural design. Excessive window spans should be avoided in areas of the house exposed to cold winter winds for as much as twice the amount of heat is lost through windows than an equivalent area of insulated wall. Generally, a total glass area of about 12 percent of the floor area of the house is adequate. In living areas, the glass area should be about 12 per cent of the floor area; in bedroom areas, this can be reduced to about 5 per cent, but at least one bedroom window should have an opener for air exchange, and be adequately sized for an emergency exit.

Energy conscious home builders must be aware of what to look for when choosing new windows for a new home. Homes are much better insulated than they were in the past, and people think that windows have kept up with this technology. However, this is not always the case. Glass is still a poor insulator. People who never had condensation in their older house now have that problem. Since insulation standards have been upgraded for the walls, attics and other parts of a house, the warm, moist air is now retained in the home, and the moisture deposited on the windows. If the house is not receiving a sufficient exchange of air during the cold months, condensation and sometimes frost will form on the inside of the window which melts, and then runs down onto the sill and into the walls causing water damage. This potential problem can be eliminated with good air circulation created by leaving the furnace fans on during the colder winter months.

When it comes time to choose your new windows, four (4) things should come to mind: the efficiency and practicality of the window, delivery schedule, and of course, the price. Unfortunately, the best windows are usually the most expensive which usually follows with practically anything we purchase, and of course, the cheapest windows are usually the least efficient.

Never has there been a greater variety of windows from which to chose. They come in all shapes, sizes and designs with different types of window materials, glazing and weather stripping. How do you select the window that is right for you? You will want to consider several criteria: your budget, the architectural style of your home, and your needs. Does the window you like give you the light, ventilation, or view that you want? Is the window well insulated and weatherproofed? Will it be easy to operate and maintain?

In order to effectively choose energy efficient windows, it is suggested that you visit several suppliers as all the brands and types available will be displayed in their show room. At first glance the windows may look very different because of the variety of sizes, shapes and different mechanical opening methods. However, the windows you choose will likely fall into 3 basic categories - swinging, fixed, or awning. Once your window shopping has been completed, the final decision on who to ask for estimates will have to be based on your own research.

Note: Before you place your order, discuss with the sales representative their policy for late delivery, and back ordering because of overbooking or broken windows while in transit. Will they repair and install the windows, or will they pay for the framer (who will have to return after he has completed his contract) to install the windows? Most manufacturers have their own service people, therefore have them include this service in their estimate so that you will not have to pay extra for something that is no fault of your own.

Window Styles:

For efficient venting, window styles such as casement or awning are chosen. Casement windows hinge on one side, either left or right, and open with a lever or crank to be held in place with a locking hinge lever. This window type can be opened just a fraction of an inch, or completely perpendicular to the house wall depending on your ventilation or cleaning requirements. (see Figure 26-27).

By installing a casement unit on either side of a fixed, center picture unit, indirect and/or direct ventilation of the room can be accomplished. For example, when the wind is blowing from the left to the right, the left casement unit is opened causing air to be blown directly into the room. If the right casement unit is opened, the exterior wind acts as a vacuum pulling the interior air outside. This is known as indirect ventilation. Because swinging windows are usually easy to operate, they can be placed in hard to reach corners, or above counter tops or dryer units.

Awning windows are very similar to casement windows except that they are hinged from the top and open from the bottom much like an awning, hence the name. Since top hinged windows direct air downward, for ventilation purposes they are best placed high in the wall, and are particularly suitable for basement installations. In areas receiving a considerable amount of rain this style of window can be left open without fear of rain entering the home which allows continual air

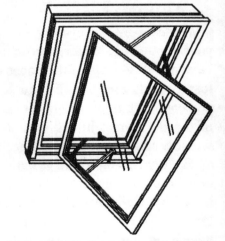

Figure 26-27. Casement window unit.

ventilation by adjusting the sash opening. Depending on the angle of the sash, an awning can also direct air upward into a room. (see Figure 26-28).

The disadvantage of the casement and awning windows is that they open out. Consequently, if the window is opening above a sidewalk or onto a balcony, it presents an obstruction to people walking by, or if on a second story it could cause a problem if someone has a fear of heights when cleaning is required. These disadvantages are minimized by the advantages of having good ventilation, an excellent weather seal, a pleasing architectural appearance, and a simplicity of opening.

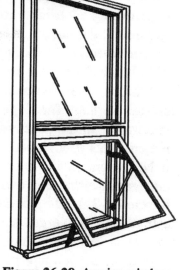

Figure 26-28. Awning window.

Window Materials:

Windows should outlast your home if they are installed and constructed properly from good quality materials. Windows may be constructed of wood, aluminum, steel, vinyl, or for improved weathering and insulation, a combination of several of these materials. Though windows with better weather protection are more expensive, they can pay off in energy savings.

1) Wood windows are warm, traditional, esthetically pleasing, and in themselves a good insulator. They are also considered the least expensive. When combining wood with vinyl or aluminum they are the best insulated, and rated low maintenance.

2) Aluminum windows being more durable than wood are also thinner, lighter, and easier to handle. When properly insulated they are energy efficient and virtually maintenance free.

3) Vinyl windows are the most durable of all the window materials, are light weight, and considered the most maintenance free of all finishes. However, with all these benefits they are the most expensive.

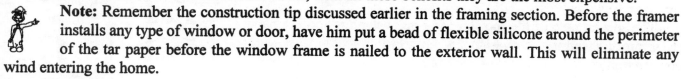

Note: Remember the construction tip discussed earlier in the framing section. Before the framer installs any type of window or door, have him put a bead of flexible silicone around the perimeter of the tar paper before the window frame is nailed to the exterior wall. This will eliminate any wind entering the home.

Window technology:

Windows, usually the weakest link in the energy system of any house, have seen dramatic increases in performance thanks to space age technology. Research that has been conducted with window coatings has shown that a reflective film when applied to the inside glass can radically improved window efficiency.

The reflective film filters sunlight of which only 50 per cent is visible to the human eye, and blocks ultraviolet and infra-red radiation both of which are invisible to humans. Radiant heat, the heat felt if you hold your hand near a hot object, is infra-red radiation. The reflective film prevents its leaving the house in winter, or from coming into the house in summer. By blocking ultraviolet light which causes furniture, fabrics, and upholstery to fade, the homeowner's investments are further protected.

Reflective windows not only chop heat loss in half, but also create additional comfort for homeowners. Because interior glass panes are significantly warmer, they prevent the cold convection currents of air that produce cold drafts, and also practically eliminate condensation. By using reflective windows not only the rate of heat loss is reduced, but the reverse is true. Because this film reduces heat build-up, the size of cooling equipment can be reduced. Therefore, significant savings on mechanical equipment can be realized using reflective window film.

◆ ROOF INSTALLER

It is the responsibility of the roofer to protect a home from the sun, snow, wind, dust, and rain via the installation of a weather-tight roof with a high durability factor. Due to the large amount of surface that is usually visible, the roofing material significantly contributes to the attractiveness of the building.

The selection of roofing materials is influenced by such factors as the initial cost, maintenance, durability, and appearance. The slope of the roof may also limit your selection of materials as low-sloped roofs for a tight weather seal require a heavier duty material than steeper roofs.(see Figure 26-29). The roofing installer estimates and charges his labor and materials by the 'square', i.e., the amount of a given type of material needed to provide 100 square feet of finished roof coverage.

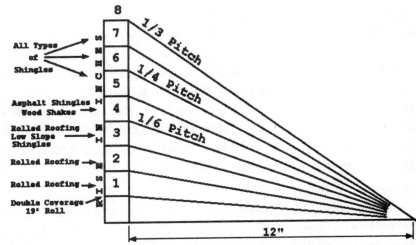

Figure 26-29. Chart showing the material requirement for a specific roof pitch.

In order to make the roof weather tight it is the contractor's job to make sure the roof is installed directly after the framer has completed the roof sheathing. Before the roofer starts to install the selected roof material, the plumber and heating subcontractors must be called to supply their weatherproof materials for venting, furnace flue, flashing and metal collars. (At the same time he will do his rough-in for the basement and main floor drains, and the plumbing stacks to the roof.) As this is being done the roofer schedules his material to be delivered and placed directly on the roof the same day. This is done for ease of access when installing, and to protect the materials against theft and water damage.

The second step of the roof preparation is stapling roofing paper to the roof sheathing in overlapping layers to protect against wind-driven rain or snow that might penetrate the roofing material. It also prevents direct contact between the roofing material and resins contained in the roof sheathing.

Areas where the roof has valleys (see Figure 26-30), or is connected to vertical walls (see Figure 16-5) must be waterproofed with metal flashing as these are the areas most likely to be affected by wind, rain, and snow if not sealed properly. Galvanized metal flashing being a non-porous material directs water away from areas of the roof affected by accumulations of water, snow and ice. The roof is then ready for installation of the selected material.

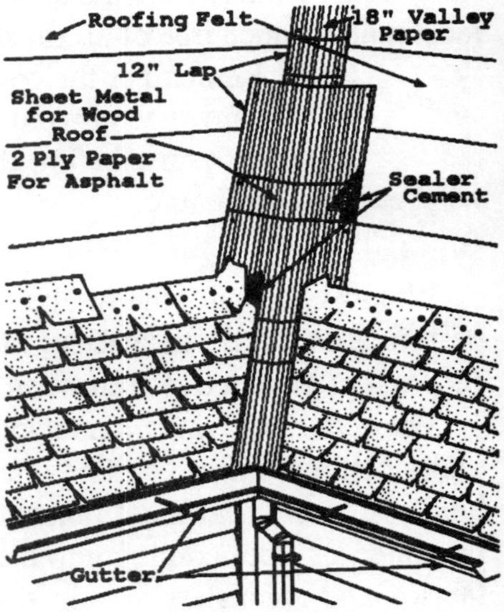

Figure 26-30. Illustrations showing flashing requirements for valleys.

Asphalt shingles are the most common roof material used by building contractors for several reasons. They come in a variety of colors, provide a good weatherproof seal, are easy to service, are light weight which eliminates excessive roof loads, and are the most cost-effective roofing material made. Asphalt shingles do not require any special structural changes to the roof trusses, and require only 3/8 inch roof sheathing for their nailing base.

A 30 pound roofing paper is laid over the sheathing material starting where the overhang and the fascia board meet. A 36 inch starter strip of roofing paper should be (but is not always) used along the eaves followed by a reversed starter strip of full length shingle laid directly on top of the paper. Both roofing paper and starter strip overhang the roof sheathing at the eaves by 1/4 to 3/8 inch to act as a drip edge for water run-off going into the gutter. The roofing paper is lapped over each strip as it repeats itself up the roof to the peak. (see Figure 26-31). Asphalt shingles are usually packaged in bundles containing 100 sq. ft. of material.

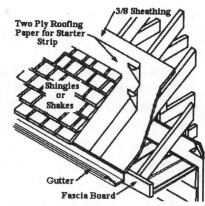

Wood shakes have been used for many years in residential construction, and are often called the aristocrat of roofing materials. Wood shakes weather to the soft, mellow, silver grey color after a years exposure that is the appearance desired by many home owners. When properly installed, they also provide a very durable roof potentially outlasting the building structure itself. Shakes must be applied to roofs with a minimum slope of 4" rise to a 12" run.

Wood shakes made from red cedar or redwood, are highly decay resistant. They are generally cut as straight split, handsplit/resawn, or taper split then graded as No.1, No.2, or No.3 utility, and available in random widths, lengths and thicknesses. (see Figure 26-32). The ends of the shakes called butt ends vary in width and thickness from 1/2 to about

Figure 26-31. Illustration showing roofing material application.

3/4 inch, and taper down to 1/4 inch or less at the opposite end with lengths of 18 or 24 inches. They are packaged in bundles, and four bundles contains sufficient shakes to cover one square (100 sq. ft.) of normal installation.

The application starts similar to asphalt shingles by placing a 36 inch strip of 30 pound roofing felt along the eaves. The beginning or starter course of the shakes is doubled. After each subsequent course length of shake is applied, an 18 inch strip of roofing paper may be applied over the top portion of the shake. This application is not always done if overlapping roofing paper had already been stapled to the roof sheathing. Individual shakes should be spaced from 1/4 to 3/8 inches apart to allow for expansion, and these spaces offset by at least 1 1/2 inches with each previous and adjacent course. (see Figure 26-33).

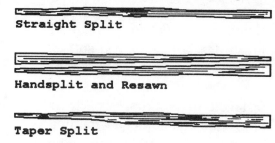

Figure 26-32. Wood shake split options.

Correct nailing is important. Use rust resistant nails or staples, preferably a hot-dipped and zinc-coated type, that are long enough for adequate penetration through the roof sheathing. Two nails or staples should be used for each shake, and driven at least one inch from each edge and about one or two inches above the butt line determined for the next course. (see Figure 26-30).

Straight-cut pine shakes have now been added to the selection of roofing materials with it's application being the same as cedar.

Concrete Roofing had been used in only a few locations across North America in the last 50 years with the product never gaining general popularity until about 20 years ago. Concrete roof tiles are now readily available throughout the United States and Canada at prices competitive with other roofing materials.

Concrete roof tiles conserve energy far better than most roofing materials. Due to the mass of the roofing tile it takes longer for the sun to heat the tile, but it retains the heat longer

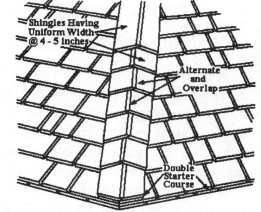

Figure 26-33. Illistration of hip application.

producing temperatures in the attic and the rest of the building that will remain cooler in the summer and warmer in the winter.

Because of the appearance, low maintenance, and energy efficiency, a building with a concrete tile roof is generally considered to have a greater resale value than one with conventional roofing, and the fire insurance rates are usually less.

Tiles come in a variety of colors and shapes from the traditional Spanish style to tiles which look like wood shakes. Compared to other materials which deteriorate and grow weaker with age, concrete actually grows stronger with the passing years. From the standpoint of longevity, roof tile is considered a lifetime roofing material. No one really knows how long concrete roof tiles will last because they have not been used that long in North America, but in Europe they have been on some buildings since the turn of the century and are still functioning well.

In areas where a heavy snow load is the norm the supporting trusses should not be less than a 4/12 slope, and must be engineered to support the additional weight of the tiles with the local calculation of roof load. The roof sheathing for concrete tiled roofs need only be 5/16" because the tiles require horizontal strapping as a nailing base. The strapping is nailed to the structural roofing trusses over roofing felt laid in both vertical and horizontal directions with a 4" overlap.

The outlookers for all roof installations are located at the gable ends of the house. It is the outlookers that provide the roof overhangs in these areas with the structural strength to support the weight of the roof sheathing, roof finish, fascia boards, and soffits. In order to support concrete roof tiles the outlookers must be fastened to the first roof truss after the gable end framing structure. The gable end, now

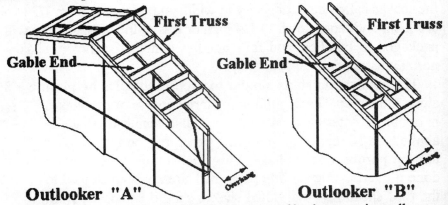

Figure 26-34. Outlooker **A** directs heavier roof loads to exterior walls. Outlooker **B** supports typical loads for asphalt roofing.

located at the mid-point of the outlookers, will be able to carry a greater amount of the direct load of the roof tiles. (see Figure 26-34). For the light weight asphalt or wood roofs, the outlooker will not be required to carry so much actual load, and can then be attached directly to the gable end framing structure.

Cleaning note: It is the responsibility of the roofer to remove any debris from the roof and around the house, and place it in the designated area for scraps and garbage when he has completed his contract.

◆ SCHEDULING CONCURRENT SUPPLIERS and SUBTRADES

Before we go into detail about the heating, plumbing and electrical in new house construction it is very important to establish a schedule so that they are not getting in each others way. Scheduling considerations start at the design phase with the main floor beams, upper floor bearing walls and floor joist layout by trying if possible to accommodate and separate the heating and plumbing lines. Early planning will help to reduce the heating and plumbing contractors cutting and removing each others lines, and thereby stop many disputes about who is responsible for fix-ups and unnecessary service calls.

If the designer located joists for each subcontractor, call the plumber in first just after the framer and before the roofer so he can complete his rough-in for basement drains and water lines before the basement floor is poured. At that time he will also run the gas line for the furnace, lines for the drains, and stacks to the roof for the upper floor(s). If the plumber is still using copper lines, it will be necessary to have him complete his rough-in before the heating contractor. This way the plumber would have first selection of joist spaces not required by the heating contractor drawings, i.e., joist spaces for the main plenum or the return air ducting. With the introduction of flexible water lines plumbers can now be scheduled after the heating contractors. The electrician is scheduled after the plumbing and heating contractors as their wiring is flexible, and can be worked around heating and plumbing lines. This will eliminate any unreported cuts or torched electrical wires.

♦ PLUMBING CONTRACTOR

It is a dirty job, but someone has to do it, and they get paid very well so do not feel too bad about them getting dirty.

Just to recap what has been discussed about plumbing in earlier chapters, the plumber is scheduled to be on the job site on several occasions. The initial visit is on the same day the footings are completed and formed. He will install the water and sewer lines just before the footings are poured so there will be no need to excavate or disturb any more ground than is necessary under the footings. They will return after the framer is completed, but before roof installation, and basement sand delivery. They will run their rough-in lines for any under-slab sump line(s), city water lines, storm sewer floor drains, basement and main/upper drains, hot/cold water lines, and all cleanouts with vent stacks to the roof. (see Figure 26-38).

Most of the plumbing pipes will run parallel to the joists hidden between the joist spaces. Sometimes it is necessary for the plumber to change the direction of some water or drain lines which will require the

drilling and cutting of studs, plates, and/or joists in order to properly line up the pipes. Where cutting and drilling is done, the structural strength of that piece of lumber is reduced.

In floor joists that are touching a bearing member, these notches or holes should be made only at the top of the joist, and measure in depth not more than 1/3 the depth of the joist, and in width not more than 1/2 the depth of the joist. (see Figure

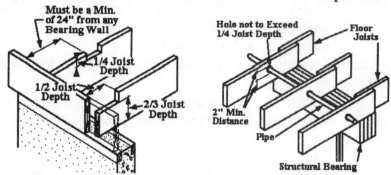

Figure 26-35. Notch and drill requirements for floor joists.

26-35). If holes or notches are required at or about mid span of a joist, the hole should again be placed no greater than 2 inches from the top of the joist, and measure no larger than 1/4 the depth of the joist (see Figure 26-35). The tongue and grooved plywood subfloor when glued and screwed properly substantially

eliminates the floor joist splitting or deflecting where the notches and holes are made. All plumbers are aware of these standards, but sometimes get carried away with their chain saw when cutting joists in order to have water/drain lines change direction. If this happens the joist will have to be structurally reinforced with plywood or joist support nailers.

The plumber will be able to do simple framing work, however, if floor joists or studs have to be moved or added, it will be necessary for the framer to return to do this work. If the depth of a load bearing stud is reduced by more than 1/3, it will require a metal stud bracket or a 2"x4" stud nailed perpendicular to it to return the member to its load bearing structural requirement. (see Figure 26- 36).

Where a non-load bearing wall has been notched or drilled, and

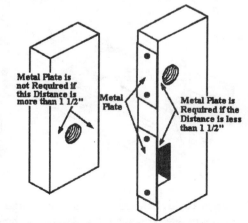

Figure 26-36. Notch and drill requirements for stud walls.

has only 1 3/4" total depth remaining, the stud will require a metal stud bracket, or a 2"x4" stud nailed perpendicular to it. (see Figure 26-36). The plumber should be held responsible for fixing these structural discrepancies, or be charged back on his final draw if the framer is called back to repair these mistakes.

At this stage of rough-in plumbing the bath tub and spa units must be placed as once the doors and drywall goes on there will not be room for them to fit. On the day the tubs are placed the insulator must first insulate and place the vapor barrier behind the tub and spa. After the units are installed it will be virtually impossible to install these items. Plumbing fixtures and accessories are not required until the cabinets, flooring, and finishing have been completed, and these items are delivered and installed at that time. This is where your over-all subcontractor scheduling must be coordinated properly.

Note: Protect the spa and tub units completely by covering and wrapping them with the same packing cardboard they arrived in. Make sure the cardboard remains wrapped around the corners, so that the subtrades when working around them do not scratch or mark the surfaces. If scratched it is very expensive to match the original color, and refinish the surface to original status.

Where any vent stacks, water line holes, or notches are cut into any top/bottom wall plates or exterior wall studs, they should be sealed with a flexible caulking or scab material to prevent air leakage from the attic, basement or outside walls. Make sure the estimate from the drywall contractor includes this in his price. (see Figure 26-37).

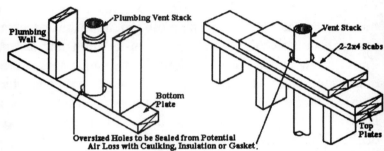

Figure 26-37. Illustration showing areas of air leakage around pipes.

It is necessary for the framer to allow space for soil stacks or large pipes which will be running perpendicular to the washroom plumbing wall and at right angles to the joists. To allow for these stacks the framer will provide headers where necessary to frame out any joists. (see Figure 26-38).

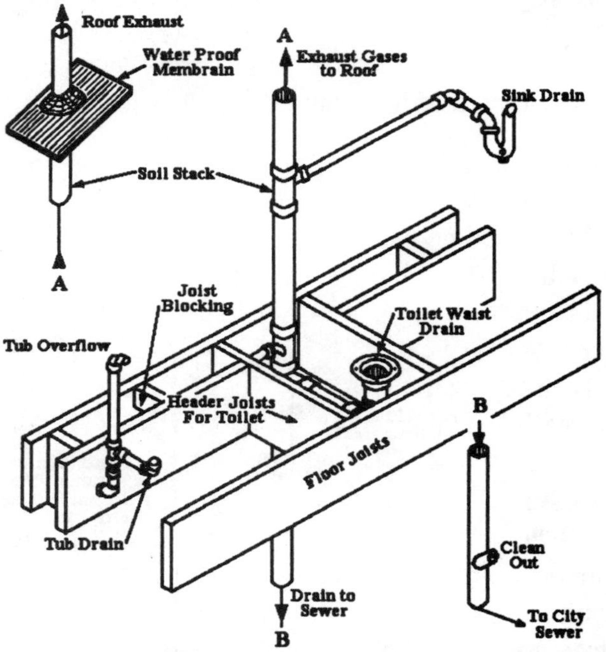

The plumber is responsible for installing the plywood gusset to which the toilet floor flange is secured. Be sure to ask the plumber doing the rough-in whether he has verified that the gusset thickness is correct for the subfloor overlay, and the selected finished floor thickness. For a linoleum floor a 1/4 inch good one side (G1S) plywood overlay should be sufficient; a tile flooring requires a more solid base so a 3/8 inch good 1 side (G1S) overlay would be installed by the finisher. If the plumber has not verified the gusset size, when the finisher installs the overlay, and the supplier tries to lay the lino or tile, the gusset thickness could be too thin or thick. (see Figure 26-39).

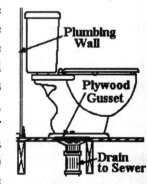

Figure 26-39. Floor gusset drawing.

Figure 26-38. Schematic illustration of framing and plumbing requirements for a typical plumbing wall.

Most Local and Federal building codes require that all under cabinet, vanity, and toilet water lines have

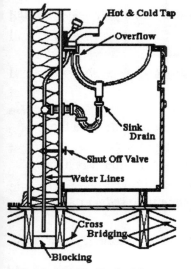

Figure 26-40. Typical location of sink shut off valve.

shut off valves located for easy access in the event of a water leak. This valve shuts off the water supply to the unit so an overflow or flooding of the floor area will hopefully be limited to a quick mop up if caught in time. These valves also provide the plumber or service person access to the unit for repairs without having to shut off the main water valve, and affecting the whole house. (see Figure 26-40). It is also a very good idea to have the plumber's estimate include shut off valves in several water lines in the basement close to the main hot and cold feeder panel. These valves can shut off lines in the basement that might weaken with age, and cause costly water damage to the developed basement ceiling, walls and floors. These valves can be turned off when away on holidays, and limit any water damage to the area of the main feeder water line rather than the entire basement. The main feeder line is or should be located above or near the sanitary floor drain.

All the drains and water lines must have direct access for servicing and repair by the plumber. In order to access these lines the framer installs the floor joists for all plumbing walls the same way. All lines are hidden in the plumbing walls, and run vertically down through the bottom wall plate and subfloor into the basement floor to the sewer. The building code requires all bearing and non-bearing walls that run parallel to the joist, and are more than 4 feet in length have double floor joists directly underneath. This however, is not possible with the plumbing walls as the floor joists under the plumbing walls must be separated to allow vertical access for the pipes contained within the plumbing walls. (see Figure 26-41).

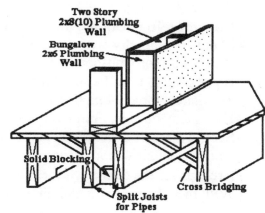

Figure 26-41. Drawing showing floor joist framing requirement under plumbing wall.

The plumber's contract includes installation of the gas lines once the furnaces are installed in the house. However, before they can install the interior gas lines, the outside meter must be installed by the utility company. The meter must be located a specific distance from any opening windows and fresh air intakes. This distance will vary from region to region so contact your utility company to find out their meter location requirements. The exterior siding, stucco, or brick should also be completed. If not, the cladding installers might not be able to install the cladding under the gas pipes and meter attached to the house.

 Note: When you call the utility company to make an appointment for site inspection, they will request that you mark the location of the gas meter with spray paint, and clean and remove all of the lumber and debris away from the working area. If this is not done to their satisfaction, you will be asked to make an appointment for another inspection, and this will continue until they are satisfied. Lack of compliance could cause several days or weeks delay depending on when another site inspection can be scheduled.

Once the utility company is satisfied with the clean-up, they will make an appointment to install the meter and the exterior gas line which can be from 7 to 21 days depending on how heavily they are booked. Once this is completed, the plumber will request a city inspection prior to installation of the interior gas lines. When approved, the plumber's gas fitter will install the rigid pipe and flexible hose to the furnace(s), and water tank or boiler. A final inspection request will be made by the plumber as a requirement of the city inspector.

As discussed earlier in chapters 12 and 25, it is necessary to provide all subcontractors with a detailed list of requirements for your house. It is critical that you discuss with the plumber the style, brand name, size and color of fixture he has included in your price estimate.

Here are a few questions you should ask your plumber about your plumbing needs:

- Are all the fixtures and colors going to match, and are all the washrooms going to have the same fixtures and colors?
- What size should the hot water tank(s) be, and should there be more than one?
- If we have two furnaces and are planning a fireplace for future basement development, does that change the size of the gas line into the house? Should we provide for that fireplace gas line now?
- Where is the best location for the gas line and meter to enter the house? Make sure that it is away from an opening window. Can they mark the best and most effective meter location?
- How many compartments should the kitchen sink have? Should it have a vegetable sprayer and a garburator, and on what side should they be?
- Should the shower heads be fixed to the wall, or should there be an adjustable telephone shower head or both?
- On which side of the sink should the dishwasher be?
- Will the main floor fireplace, garage space heater, deck barbecue, kitchen appliances, or clothes dryer require a gas line?
- Should there be a small bar sink placed somewhere for children?
- Does the refrigerator require a water line for ice or water?
- Should the laundry area or garage have a sink or tub for clean up?
- Will the spa or whirlpool tub have grab bars and sufficient jets correctly located?
- If planning to install a sprinkler system at a later date when affordable, should we now make provisions for a 3/4 inch water line in the basement?
- How many outside, nonfreeze water taps should there be for proper coverage?
- Should there be a water line to the garage or basement? Should it be hot and cold?
- Does the laundry room require a floor drain?
- Are taps for the clothes washer included in the estimate?
- Will there be rough-in plumbing located in the basement, and should it be for a 3 or 4 piece bath?
- Does our area require a sump pump? If so, does his quote include intake and exhaust water lines, and installation of the sump pit and pump? Will it have a reverse back water valve to eliminate back wash?
- Do we require an in-line back water valve for the basement floor drain to eliminate back-ups and basement flooding?
- Have you discussed shut off valves for water lines in the basement?
- Should we make provisions for any future water lines for a steam shower or sauna?
- Will a water softener be required to save on water consumption, and minimize cleaning of the humidifier unit?
- Do you require a floor drain in the garage, and will the installation of the pipe and grill be included?
- Will the plumber insulate and caulk around all exterior pipes and holes to stop air infiltration and freeze ups? See Figure 26-52 on page 133.

♦ HEATING CONTRACTOR

As discussed in previous chapters, it is the responsibility of the designer to provide drawings which accommodate for the main heating plenum, and the room ducting and return air for the basement, main, and upper floor joist cavities. The simpler the house design the better the heating efficiency.

The heating contractor's first contact with the house is when he drops off the premeasured furnace flue flashing, chimney collar, and bathroom fan venting for the roofer to install. You will see them nailed to the front wall of the garage or house to be seen easily by the roofer.

Have the heating installer include in his estimate the supply and installation of the sheet metal cap for any stick-framed chimney chase. This metal cap is made with a slope to shed any water away from the furnace or fireplace flue pipes. The heating contractor will be responsible for the installation and water-proofing of the storm collar for the furnace flue; the fireplace installer will be responsible for the installation of the collar to the fireplace flue.

Note: If not clearly stated the furnace contractor may not supply and install the metal chase cap. He will assume the fireplace contractor is responsible, and the fireplace installer will assume the reverse is true. Also, find out the fireplace flue diameter so that the furnace contractor can cut the correct hole size for a proper fit. As you can see this is one of those areas that overlaps between contractors and needs to be controlled.

The heating contractor provides the bathroom exhaust vents for the bathroom fans, but there are alternate venting methods. To believe that a house is energy efficient, you first have to be satisfied that as little heat as possible is radiating up fan ducts to the roof. With fewer holes in the roof there is also less chance of water leakage especially when the house gets older. If there are no drafts blowing down through a bathroom fan with a bad damper, there will be less indirect loss of heat through the roof. If no roof vents are sticking out, the roof lines look cleaner, and you save about 15 dollars worth of sheet metal and a few dollars in heating costs. How is this done? Very simply, have the heating trades reroute the flexible fan vent to the soffits. This way the exhaust air blows out the soffit, and any wind that is blowing in is stopped by the soffit before it hits the vent. Wind blowing into the bathroom and cooling the air is virtually eliminated therefore creating more comfortable space and energy savings.

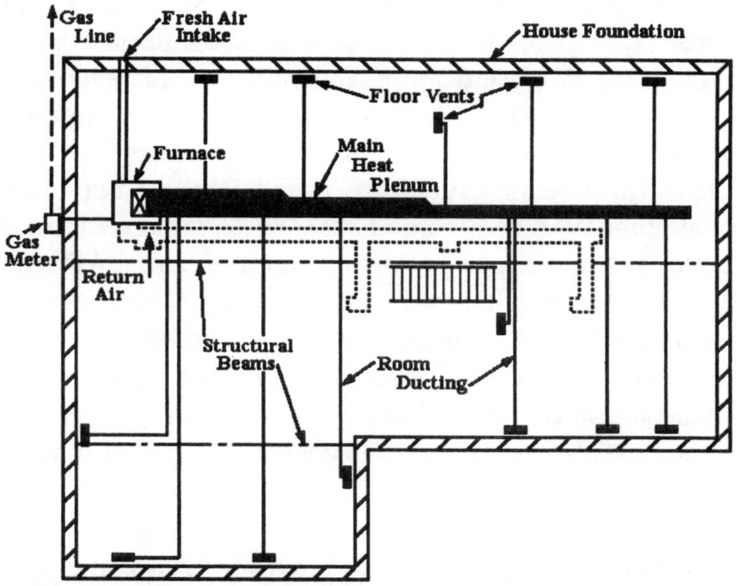

Figure 26-42. Plan view of a homes typical heat and return air ducting.

Forced Air Heating:

A good heating contractor should be able to supply a B.T.U loss calculation and a schematic drawing of the heating and return air ducts for your house, so make sure you get it in writing that this will be provided when you accept the estimate. This B.T.U loss calculation and schematic drawing is required by some financial institutions as part of their mortgage package. (see Figure 26-42). It will also tell you that the heating contractor knows how to read the blueprints, and is up to date with his heating systems. When reviewing the schematic drawing with your heating contractor make sure that return air ducts are provided for all hallways and rooms. They are not required in areas where people do not spend

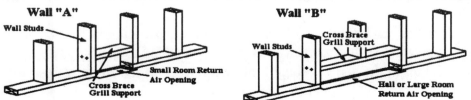

Figure 26-43. Return air wall framing requirements. "A" - Small room. "B" - Large room.

much time such as bedroom walk-in closets, bathrooms, and storage closets. These areas are always considered when calculating the total return air requirements for the adjacent rooms or halls. (see Figure 26-43).

Whether you heat with oil or natural gas the energy efficiency of furnaces will vary from low to mid to high. The more efficient the furnace, the more complicated and the more expensive it will be. Mid-efficient furnaces with electronic ignitions are the most common, mainly based on price, and are rated from 64 to 68% efficient. These furnaces require a vertical metal chimney flue to exhaust the heat and furnace gases to the roof. These metal flues become very hot, and must be separated at least 2" from any wood structure by using a sheet metal flue shield. These shields should be installed by the heating contractor during installation of the furnace ducting, and prior to installation of the wall insulation and vapor barrier. (see Figure 26-44).

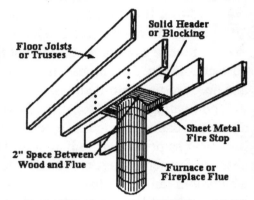

Figure 26-44. Location of the flue shield.

The high-efficient furnaces, which can be from 96 to 98% efficient, require only a side vent system to exhaust the furnace gases. These exhaust pipes are sometimes made of a heat resistant plastic which lead directly to an outside wall of the house. One 120,000 B.T.U. high efficient furnace will heat an average sized house without any problems. However, this creates a single control or zone which heats the entire house, i.e., the basement, living, family and bedroom environments, at the same time. In order to best divide the heat equally throughout the house, the furnace needs to be centrally located, [but, this is not always possible depending on planned basement development]. (see Figure 26-42). To accommodate for different family life styles, heat in a room can be reduced or increased by adjusting the floor vents. After living in a house for one year you will know which rooms to adjust, and how much to reduce or increase the venting.

Where living environments in houses such as large bungalows and two stories are further apart, consider having two furnaces and therefore two zones. One furnace could heat the living, family and kitchen areas along with their counterparts in the basement, and the second furnace heat the bedrooms and bathrooms along with their basement counterparts. You can still reduce or increase the amount of warm air entering each room by adjusting the floor vents.

There are a few convenient options that you might want to have installed on the furnace to reduce on-going maintenance functions. Most furnaces come equipped with small, inadequate humidifier units, which have to be upgraded in order to allow for sufficient humidity in the house. The most common upgrade is to the drum humidifier which will provide sufficient humidity, but will have to be maintained every six months by dismantling and cleaning the sponge rotation drum. The build-up of mineral deposits left by chemicals added to the water must be routinely removed as too much residue will strain the motor causing burn out. Many home owners are selecting a more expensive but efficient drip humidifier which supplies more than enough humidity directly into the house via a constant water drip into the main heating plenum. Maintenance will depend on the degree of mineral deposit build-up in the plenum from the hard water. With all humidifiers almost all the mineral deposits are eliminated with the installation of a water softener by the plumber.

Another good option that can be installed to keep down dust, particularly for people with allergies to animal fur or pollen, is an electric dust zapper. This unit when added to the furnace electrically zaps dust particles that are circulating in the air, or brought into the house via the furnace fresh air intake or open doors.

If a garage space heater is required, it is hung about 8 inches below the ceiling at the back of the garage, and faces the front garage doors. This direction is best suited to prevent overheating the fire rated wall separating the garage and the house. Space heaters are manufactured in several sizes varying from a 40,000 B.T.U unit suitable for a well insulated, double garage to a 75,000 B.T.U unit capable of heating a well insulated, triple car garage. There are other sizes available, and your selection will depend on the type of work you will be doing in the garage. Discuss this with your heating contractor so he can best determine what your heating requirements will be. The installation of the unit heater and its vent flue will be completed by the heating contractor prior to the plumber connecting the gas line.

 Note: When designing your heating specifications have the subcontractor make allowances for future basement development by running ducts to the windows for each room that will require heat. Until the

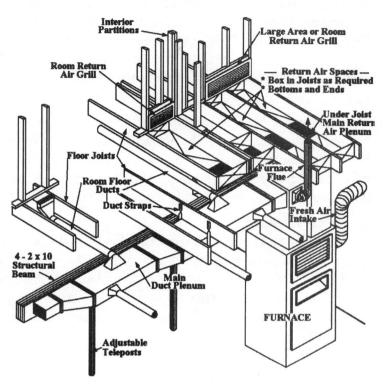

Figure 26-45. Schematic drawing of a typical forced air system.

basement is fully developed you can reduce the loss of heat through these ducts by closing off the dampers located inside and adjacent to the main plenum. Also make sure the heating contractor includes the basement dampers in his estimate.

All types of heating systems can be safely and easily installed in a wood framed house. There are however certain fire separation devices that have to be used, and clearances that must be maintained between parts of the heating system and combustible wood materials. Your heating contractor will know these building regulations, but for safety's sake request that the city building inspector check these areas of concern after the heating contractor has completed putting in the furnace flue, and prior to installation of the insulation. When doing visual inspections, there should be a minimum 2 inch clearance provided by metal spacers, shields or strapping between any flammable material and hot surfaces.

It is the responsibility of the heating contractor to supply and install any sheet metal work or venting ducts required for items such as the cook top fan, laundry dryer, and hot water tank.

 Note: The framer when on-site cuts any holes for vents that go to an outside wall. Have him cut the hole as close as possible to the duct size which will be provided to him by the heating contractor. Once installed the heating contractor will seal around the perimeter of the vent with insulation and caulking to eliminate potential air leaks or rodent entry.

Hot Water Heating:

This heating system has been around for many years, but only recently have sufficient changes been made to allow this method to become more affordable. The basic principle of heating water and sending the heated water through pipes remains the same, however, the use of old steel wall radiators are outdated. Flexible plastic tubing capable of handling the necessary hot water temperature is now being used to heat the total floor area of a room with each individual room having its own zone control. These flexible tubes can be placed within concrete floors, or under wood floor structures by having them run parallel along the floor joists and secured with brackets to the underside of the subfloor. This will provide sufficient heat to warm the floor, and keep the radiant water temperature comfortable to the touch without causing damage to any adjoining wood materials. Separate ducts for air circulation will be required to eliminate high humidity levels, and the subsequent condensation on windows resulting from radiant heat. Hot water heating systems are initially more expensive to install, but are very energy efficient. They will justify themselves with cheaper monthly heating bills, and provide your family with years of comfort.

Hot Water/Forced Air Heating:

A hot water heating system can be combined with a forced air system to bring the benefits of both heating systems together. The hot water is heated by circulating in coils contained within the boiler. Blown air is passed over the coils, heated, and then sent through a duct system similar to a forced air system to each individual room. The design of the heating system allows each room to have its own thermostat therefore providing each family member the control over their own living environment.

Electrical Heating:

Since electrical wires are easily hidden behind stud walls and under floor joists, the planning and design specifications for electrical heating systems are negligible compared to that of forced air and hot water heating. Electrical heating units are located around the perimeter walls of rooms and under the windows to allow heat to radiate evenly throughout the rooms using the convection principles. Because the heating units are mounted against the inside walls, there is no need for cutting and blocking the floor joists and wall studs.

Electrical heating in areas where natural gas and oil are abundant is not economical, but where it is too expensive to run underground gas lines electrical heating is usually the only alternative system. An electrical heating system combined with a forced air system will bring the same benefits of both systems together. The air is heated by blowing over hot coils, then circulated through a duct system similar to a forced air system to each individual room. The design of this heating system, similar to a hot water system, allows each room to have its own thermostat.

With more energy efficient houses being built it is necessary to add additional fresh air to the rooms using a separate ducting system. This added air helps to eliminate the condensation usually caused by slow, electrical radiant convection.

◆ ELECTRICAL CONTRACTOR

The electrician is first called to supply temporary power from the meter to the house if the subcontractors do not have portable power generators, or there is no access to a neighbor's power, i.e., unable to purchase temporary power from one of their outside plugs. Prior to connecting the temporary power from the meter, you must have the underground power line and meter installed and inspected by the city. This installation can be done before or after backfilling the foundation. (Remember to make sure that the foundation and weeping tile inspections have been done by the city before the backfill.) Some electrical contractors are able to supply this service if requested. Otherwise you will have to get another contractor to supply these services, and the electrician should be able to supply some names. If the meter and the subfloor are installed at this point of the construction, the electrician can install the main electrical panel. Make sure that the location of the main panel is clearly marked so that the electrician will not install the panel in a future developable area of the basement.

The electrical wiring is usually started when the house is completely water tight, i.e., the roof completed, windows installed, and plumbing and heating rough-in completed. This phase of wiring, usually called rough-in, includes the drilling of holes in the studs for wires, and installing the wiring to the electrical boxes for wall switches and plugs, and octagonal boxes in walls or ceilings for light fixtures. The design and installation requirements are usually controlled by the Local or Federal building codes, and are inspected by the city building inspector to ensure compliance with the electrical codes.

Because the location of switches, plugs, and lights is so important, discuss with your designer in detail all the planned locations of the appliances, televisions, telephones, table lamps, etc. As new products become available and existing products have more features added to them, it will be necessary to consider alternatives for the electrical design of your house. Home computers, portable rechargeable vacuums, and built-in vacuum power heads all require special considerations. Computers require several wall plugs in close proximity to each other so that the computer terminal, printer, and modems are close by without requiring extension power bars. For quick access a rechargeable vacuum on a wall location requires a wall plug of its own to provide constant recharging after use. Consider the convenience of a unit in the hallway for the bedroom areas, and another in the kitchen for the living areas of the house. Built-in vacuum systems with power heads for deep carpet cleaning require their own power source located within a few feet of the vacuum suction vent.

If you recall in chapters 3 and 16 when you visualized a building on your lot, you did not start with a preconceived idea of the house you would like to build, rather one that fits your basic life style. To visualize the electrical requirements in your home, walk through the house room by room in your mind as you would normally do when living in it day to day. Picture where the switches should be located in conjunction to the door swings, and how furniture arrangements will determine the best locations for wall plugs and ceiling light fixtures. Will some switches control any outside security lighting, or interior wall plugs for table or floor lamps? Should you have any 3 or 4 way switches located in different areas of the house controlling one light fixture? An example of a "multiple switch" is a wall switch located near the main entrance door and another switch located near the laundry room door which control a single hall ceiling light. This is called a 3-way switch, but if you wish a 4-way, simply place another switch for example at the bottom of the stairs to control that same hall light. All basement stairways and stairs leading to a second floor should be controlled by 3-way switches. Also discuss with your electrician the color and style of switch and plug plates as they should match or coordinate with your interior design color selections.

Electrical requirements for special lighting such as exterior Christmas lighting should be considered at the design stage of your house. If you plan on stringing lights on your eaves trough consider soffit plugs located at several corners of the house with interior control switches. Special water-proof plugs for exterior spot lights and tree lights can also be installed and controlled from inside the house. Your designer can make recommendations as to their locations so that all electrical requirements are incorporated into the design, and registered on the blueprints with notations. These details will help to ensure an all-inclusive estimate by the electrical contractor. Consider a dimmer switch to control the amount of light required for different occasions, eg., family room track lights for bright lighting, or mood lighting. If you are considering ceiling fans in some rooms to circulate heat in the winter or cool the room in the summer, then make sure you have the electrician install a rheostat to control the low, medium and high speeds of the fan. This eliminates having to pull the fan's chain to change the speed or dragging out the ladder if the fan is located on a high ceiling.

Once the number of switches, plugs, and lights required has been determined, make sure that the electrician has provided a large enough service panel with sufficient breakers for present needs as well as additional empty breakers for any future basement development or renovations. This prevents later having to add an expensive and poorly located sub-panel. It is illegal to double up on any existing breakers as this practice will most likely overload that circuit, and possibly cause a fire. Most underground city services are capable of providing a 100 amp./240 volt service to the house. This is sufficient power to provide you with a 48 or 64 circuit main panel which is large enough to handle all present and future electrical needs. Locate it in the basement, garage, or on the main floor based on ease of access. You might wish to split the panels by placing the main panel of 48 circuits in the basement, and a secondary sub-panel of 32 circuits in the garage or on the main floor. The sub-panel could contain the heavy duty circuits such as range, washer, dryer, spa tub pump, cook top, etc., and provide easier access to a circuit which needs resetting.

The following information should provide you with a list of detailed specifications to be included in the electricians estimate for your house.

- Total number of light, plug and switch outlets.
- Underground service provided - No. of Amps and No. of Volts.
- Will you require wiring for a 220 volt plug in the garage for power tools?
- Is the underground trenching and wire included?
- Number of circuits in the main panel to be 48 or 64, and where will it be located?
- Is a sub-panel required in the garage or main floor, and how large should it be?
- Wire, install the oven, and connect the power supply to the oven(s) - (30 AMP./240 Volt).
- Wire, install the cook top, and connect the power supply to the built-in cook top - (30 AMP./240 Volt).
- Install dryer plug - (30 AMP./240 Volt).
- Wire and connect furnace(s).
- Wire and connect humidifier(s).
- Supply, wire, and install smoke detector(s).
- Wire and connect dishwasher, garberator, and trash compactor.
- Wiring for water meter and gas meter.
- Supply and install perimeter motion sensor lights with interior control switches.
- Wiring for telephone outlets - complete with jacks
- Wiring for cable television outlets - complete with jacks.
- Supply power for intercom.
- Supply power and install door bell chime.
- Wiring for garage door opener(s) and push-buttons.
- Install exterior ground fault protected waterproof plugs/plates.
- Supply, wire, and connect bathroom fan(s) No. _____.
- Supply all ground fault protected bathroom plugs.
- Wire and connect spa tub pump (15 AMP./120 Volt)
- Wire and connect hot tub and pump switches, steam shower unit, sauna heater, etc.,
- Install plug for main vacuum supply unit.
- Supply and install standard (color) plugs and switches.
- Supply and install sliding dimmer switches - # single pole and # 3-way sliding dimmer switches.
- Wire and install front yard light standard, trenching included (light unit supplied by customer).
- Wiring for garage space heater unit.
- Hang # owner supplied ceiling fans, and supply and install # rheostats.
- Supply and install # insulated pot light cans.
- Supply and install # standard pot light cans.
- Supply and install # - ? watt (color) open potlight trims.
- Supply and install # - ? watt (color) eyeball trims.
- Supply and install # heatlamp cans complete with trim.
- Build # attic boxes around heatlamps or as per code.
- Install sump pump plug.
- Supply and install # - # tube florescent trim/wrap units.
- Supply and install # - under cabinet # tube # foot side mounted florescent.
- Supply and install # - # tube # foot florescent strips for sunshine ceiling.

- Supply and install 15 AMP. 240 Volt plug(s) in garage.
- Supply and install poly vapor hats around electrical outlets on all outside walls and cold ceilings. Look for this during construction to eliminate any costly mistakes.
- Hanging and assembly of all fixtures.

It is important to provide as much information as possible to the electrician when selecting electrical needs. When the electrician is ready, make sure that you have the built-in ovens, cook top and dishwasher delivered to the job site A.S.A.P. so they can be wired and installed into their positions. The rough openings in the kitchen counter top, shelves and floor, and cabinet base for the cook top and down draft venting are precut by the kitchen cabinet installer. Supply the cabinet installer with the correct measurements or templates to properly cut the correct rough opening size. The dishwasher's flexible water lines are connected by the plumber before the electrician connects the power, therefore, the plumber must allow sufficient play in the lines for the electrician to move the dishwasher in and out.

Depending upon the electrical contractor, many can also supply an estimate for wiring and installing vacuum, intercom, and security systems. However, get additional estimates from other suppliers who deal specifically with these products for a price comparison as you might find them to be cheaper. These products will be discussed in more detail in this chapter.

♦ BUILT - IN VACUUM SYSTEM

The vacuum installer will require access to the house after the heating and plumbing contractors have completed their rough-ins, but prior to the electrician running his rough-in lines. The vacuum system selected will vary depending on the size/length of hose, number of service inlet locations, total length of the main collection lines, and the options available with the unit.

When collecting estimates for the system ask the following questions:

1) - Are there quick release fasteners for easy access to the bags or collection cannister containing the dust?

2) - Is there an automatic shut off switch in case of overheating?

3) - Does the vacuum unit include options for service inlets for easier cleaning of basement, or garage, workshop and car?

4) - Will the proposed length of hose be sufficient to reach all corners of the house from the selected service inlets on the plans without moving furniture or over-stretching the hose line?

5) - Is the motor's suction strong enough to pick up soil and dirt from deep in the pile of the carpet using either the standard or the power vacuum head?

Discuss and confirm the cannister and service inlet locations shown on the working drawings with the supplier prior to installing the system. You must first know where the cannister is to be placed so that the installer can run his main collection line to it, and the electrician can install the power plug. If located in the basement, will it interfere with any future development, or be too noisy to carry on a conversation if the vacuum is running? Most importantly, will you have easy access to it for cleaning and servicing? Because of these potential problems, many home builders suggest that it be located in the garage, i.e., away from living areas where the common wall between the house the garage acts as a noise buffer during operation. The garage location is usually more accessible for servicing, closer to the garbage area for emptying, and eliminates carrying a full cannister up stairs.

After installation of the cannister, the light weight plastic pipes are installed between the floor joists from all the separate service inlet locations, and connected to a main collection line running alongside a structural basement beam or bearing wall. Power heads are usually considered an option, however, in order for deep seated dirt to be removed the power head may be necessary. Finally, the color of the face plate for each service area should match or complement the color of the face plates for the switches and wall plugs.

♦ INTERCOM SYSTEM

The technology for intercom speakers has not changed all that much in the last ten years, but the attachments and features have advanced with the times. In addition to the usual AM and FM radio options, and the interaction with the separate room units, they can also include a tape cassette, CD laser disk, hands-free operation, and on/off timer features. If money is no object another option is a television monitor to the front door, but remember, the more features and speaker controls the greater the expense involved.

Rough-in for the intercom system can be done during the framing stage at the same time the vacuum and security systems are being installed because none of the lines will interfere with each other. Once again the locations of each individual speaker and main control panel are pre-determined by yourself, and shown on the working drawings by the designer. When discussing the placement of the speakers and master control panel, be sure that you have considered where the electrical switches, plugs, and most importantly furniture will be located in each room. There is nothing worse than having a speaker or master panel behind furniture or a picture.

The location of the master panel should be central to the house, and in an area of frequent use by most of the family to allow monitoring of rooms and the front door. The master control panel is best situated in the kitchen, near a telephone desk, or in a family room wall that is not visible from the front door. When locating the individual speakers, you should be aware that speakers, switches, and wall plugs on the same wall within 4 feet of each other, and/or on the back side of the opposing wall will cause static in the lines, and result in poor quality sound. Sometimes optimum spacing is not always possible, so try to separate them as far as possible on different walls.

Consider the following locations for individual intercom units - bedrooms, bathrooms, dining/living room, family room, garage work area, outside deck or patio, basement, laundry room, front door, and office or den areas are all good locations for music and monitoring the front door. If you have considered placing a speaker in a bathroom, review your Local building code, or ask the city's building inspector for its possible location with respect to any water hazards. You might need an exterior water-proof speaker for this area.

Light switches are usually located as close to the opening side of a doorway as possible with the speaker beside it. Because the framing of the studs are 16" apart, if you are going to place the light switches on the closest stud to the door, placing the speaker on the next closest stud (approx. 16" away) might be locating it too far from the door or too far into the room. You might have to consider placing the speaker on another wall, or above or below the light switch. Depending on the room, locate the speaker where it would be convenient to reach, eg., in a bedroom it could be over an end table, or on an otherwise unusable wall.

♦ SECURITY SYSTEM

Many people say that a house is the single largest investment that a person or family will make in their lifetime, so it only seems appropriate that that investment be protected by a good security system. The system that you select will depend on the degree of comfort and safety you wish to feel in the home with the locking devises installed on your doors and windows, and the extent to which you wish to protect your personal property and family against potential break and enter.

Is your neighborhood considered high risk because it is a new subdivision without fences? Does it have poor street lighting, and back alleys, or is it just a middle or upper income subdivision which brands itself as "quality pickings" for the break and enter types? Do members of your family leave for work or school at the same time, and return at approximately at the same time every day? Do you like to go out frequently for dinner, to a movie, or visit friends? Do you have a color television, radio, CD player, video recorder, jewelry, or limited edition prints? If you answered yes to any or all of the above, then you should consider security system to protect your property.

The best, noisiest and least complex, however, less economical or maintenance-free answer is a dog. The most practical and least expensive electronic security system can be as simple as lamps plugged into timers, and perimeter, motion-sensitive floodlighting; the more high-tech, micro-chip, master panel controlled systems comprising pressure-sensitive floor plates, glass break and motion detectors, and door contacts are directly dependent on your allocated budget. Sirens are usually installed in the overhang at the front of a house, and the return air plenum which acts as an echo chamber for the siren.

To choose the type of system best suited to you, discuss your fears or concerns with the security sales representative, and review the house blueprints and proposed landscape plans. Try to visualize the potential traffic pattern of the burglar, and protect those areas with the correct type of detector. Take into consideration either the family being away, or someone being alone in the house. Burglars like to select break-in areas that are dark, and hidden by a fence, tree, garden shed, or object to avoid being seen. If glass is to be broken to gain access, the smaller the area of glass the less noise it will make. Consider the following patterns of movement through the house.

1. Basement windows are usually the most accessible, low to the ground, and usually dark.
   Basement access reduces noise caused by furniture or items being tripped over and/or broken.
2. Once entrance to the basement is gained, they have direct access to the upper floors via the stairs.
3. Once access to the main floor is achieved, the hallways will provide direct routes to the valuable rooms of the house.
4. Entry via a main floor break-in would probably be in an area that is dark, and hidden by a tree or part of the house. Do any of these areas have easily accessible windows, or french or patio doors from a deck or patio with poor locking devices? Check for quick entrance and exit possibilities.
5. The most desirable target areas are usually the family room for the electrical equipment, dining and living room for the art and valuable display items, master bedroom for the jewelry and/or money, secondary bedrooms for portable electronics, and the kitchen for any money or just pure vandalism.

Having identified the areas of concern, it is your representative's job to help identify the areas to be secured, and select and locate the key pads and best sensors. Security keypads are usually located near a door that is used by most family members such as the mandoor from the garage to the house, mud room access, or the front entry. It is also suggested that another keypad be placed in the master bedroom or bedroom hallway for access if someone is home alone and arms the security system. All security systems can be installed so that a family member alone in the house can roam about without setting the system off. However, once a secured door is opened the security will be activated, and the sirens will sound. Security systems work on the principle that when a person decides to leave, he/she enters a code into the keypad activating the system. There is a given period of time in which to open the door and leave, and once the time period lapses the security system will arm itself until someone re-enters the house, and provides the correct code to disarm the system.

Here are a few suggestions of the types of detectors available to secure different areas of the house.

A (protect perimeter of house). The house perimeter is best and most economically secured with several properly located, inexpensive flood lights controlled with motion sensors. These lights should also be controlled with an interior wall switch.
B (protect basement entry). The basement windows and basement itself should be controlled with at least two different types of detectors. If burglars can access through a basement window, magnetic pads can be placed on the opening part of the window. In case of breakage of the non-opening part the window a glass break could be used, or if the basement is undeveloped a single motion detector

could sound the alarm. If the basement is to be developed any combination of the above can be used for different living areas of the basement.

C (protect stair access). If for some unforeseen reason basement access is not detected, a secured stairwell will prevent burglars from continuing. Either a motion detector at the top or bottom of the stairs, or if there is a door at the top or bottom, a magnetic door pad contact. Once the door is opened the alarm will sound.

D (protect hall access). The hall whether on the main floor or upper floor provides access to all the rooms of the house. Since there are usually no windows in a hallway, they are best secured with a motion detector located at one end of a hall, and aimed so that any motion between 18 to 60 inches up from the floor will register as an intrusion.

E (protect potential access points). All possible intrusion areas can be covered using one or more security devices. If an area has no doors a glass break will most likely be the security device of choice. Where door access is possible through a french door in a kitchen, mud entry, or family area, these areas can be protected with door contacts. It is possible with these door styles to gain access by breaking the glass alone, therefore a motion detector correctly located could secure several rooms at one time, and eliminate the need for several glass break units.

F (protect personal valuables). Each room should be protected differently depending on the possible access route and the valuables contained. Large rooms like master bedrooms can be best secured with motion detectors located at a corner point of the room. Small bedrooms can be secured using glass break. If the glass break fails to catch the burglar for any reason the hall motion detectors will activate the system.

Your security systems and smoke detectors can be connected to the local fire and police stations in addition to being monitored by a security company. For a small monthly fee they monitor your security system from their office, notify the appropriate municipal department should something happen, and send one of their security people to the house to let them in. If you are on holidays, they will notify a relative or friend so that the house can be serviced or repaired. Monitored systems increase the sense of security to family members when they are alone at home, or away from home on business or holidays.

Installing your system at the correct time during the construction period is essential to make sure that workers cannot mistakenly disconnect, cut, or cause damage to the wiring. The security installation should be after the electrician has completed his wiring, and just before the insulator/drywaller starts. If you cannot afford the full cost of the security system, many installers will provide the rough-in wiring at a reasonable price allowing you to complete the system when you can afford it.

♦ CHOOSING YOUR LIGHTING:

There are three important considerations when selecting lighting requirements: decor, energy costs and lighting needs. Good lighting brings a room to life. Lighting systems should provide a visually comfortable and safe level of light for the family's activities, and accentuate the elements of your decor using general lighting, local lighting, and accent lighting techniques.

General Lighting encompasses the total light available in a room needed for seeing, housekeeping, television viewing, etc. It must be sufficient to ensure safety and allow the performance of simple tasks. It must minimize eye fatigue by reducing the contrasts between selective lighting and all-around lighting. In order to adequately light an entire area, space recessed fixtures evenly so that the distance between them is not more than the distance from the top of the work area to the ceiling. Around the room perimeter they should not be more than 36" away from the walls, or 18" away from the edge of any furniture placed against the walls. Recessed fixtures use a substantial amount of bulb wattage which can be minimized by using sliding dimmer switches to save energy, extend the bulb life, and allow adjusted lighting levels in the room for various occasions.

Local Lighting is provided by lamps or small fixtures. Local lighting raises the level of the room's general lighting requirements to fill specific needs for selected areas, eg., a reading lamp next to a bed.

Accent Lighting focuses on an interesting object in a room and enhances the colors in the furniture and artwork, or the texture of the walls themselves. Tracklights, potlights, and eyeball lights are examples of lighting used to gain this effect.

The balance of these three types of lighting, their interaction with natural light, and the selection of fixtures and lamps determine the total effect of any lighting system. Efficient and effective lighting requires careful planning. In order to achieve this make a daytime and nighttime inventory of all the rooms lighting needs. First visualize the amount of sunlight that could enter the rooms on a sunny day, and then determine the minimum of daytime lighting in those rooms, both in winter and summer. Then identify nighttime activities in each room, and select the lighting required for visual comfort. Finally, decide the mood you wish to create in each room, i.e., restful and serene, or dramatic and lively. The lighting sales representative should be able to assist you with lighting selections given the positioning of the house on the lot, and relation to the North - South sun exposures.

Task Lighting is the amount of lighting you require to comfortably perform specific activities such as cooking, reading, sewing, shaving, applying make-up, studying, and hobbies to prevent eye strain and accidents. Do not work in your own shadow! Direct the light to the applicable work, play and study areas. Install fixtures to the sides or back edge of work surfaces to avoid shadows or glaring reflections. Consider everyone in your home and the location for various activities as most areas need different lighting suitable for a variety of activities.

Bedrooms: Each bedroom can be a study, sewing room, television room, sickroom, or games room. You need good lighting for seeing in drawers and closets. For reading in bed, consider a "hanging lamp" which saves table space, track lights, or wall lights. Young children often play, study, or read at a desk, on the bed, or on the floor of their bedrooms. These areas therefore need high level lighting.

Kitchens: Fixtures should be located directly over work areas to avoid shadows on the work surface. Mount fixtures as close to the center of the sink or island as possible. Dropped florescent island, or complete sunshine ceiling units are used for high level, non-glare lighting. For work areas, use florescent lighting under the front edge of the upper cabinets. For the dinette area, a decorative hanging fixture, or a fan and light fixture combination is centered over the table.

Living/Family areas: Many designers and custom home builders favor recessed lighting for the look of clean, contemporary ceilings, but will use recessed lighting with traditional furnishings too. Recessed fixtures are especially useful for lighting corners where people do not wish to use traditional chain swag lamps, or low-ceiling areas such as basement recreation rooms. There are many styles and sizes of recessed fixtures available. The open potlight generally gives the most illumination; eyeballs project below the ceiling line and can be directed at any area or object. The general lighting of a beautiful chandelier can be supplemented by functional and dramatic recessed potlights over dining and server areas.

Lighting can be an important part of your decorating scheme. For example, use pot lights or eyeball lights to make a small room seem larger by accenting the walls with light. Use dimmer controls to make a large room seem cozier, or softly light a conversation area. Use eyeballs to dramatize the texture of brick, stone, wood and draperies, or spotlight bookshelves, sculpture, prize prints and wall hangings. Show off plants with an interplay of light and shadow.

Lighting walls can create visual interest, make a room seem larger and improve the general lighting level. Choose pot lights or eyeballs according to the style, wattage and the area or space to be lit. Eyeballs tend to give a reflective lighting effect at the ceiling line which can be reduced by either placing the eyeballs closer together, or requesting a non-reflecting inner baffle with the trim kit selected.

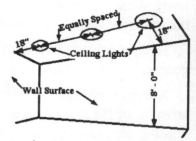

Figure 26-46. Drawing showing ceiling eyeball locations.

To accent the dimensions of textured walls and draperies, install recessed fixtures closer to the wall; to minimize reflected glare from glass covered artwork, install an adjustable eyeball slightly to the side of the picture. Adjustable eyeballs are also excellent for lighting sculpture and plants. These objects are often best displayed with back, side or front lighting rather than direct overhead fixtures.

Built-in lighting is expensive, which may make it too costly and impractical. The effect of built-in eyeballs and potlights can be simulated with spotlights attached to tracks mounted on walls or ceilings. A track light system can have different styles, and any number of lights that swivel at selected angles for either general illumination, or spotlighting specific areas or objects. Create interesting effects by using different types of bulbs on two or more tracks set in a parallel, perpendicular, U-shape, or rectangular arrangement on the ceiling. The new halogen bulbs available for track light units can also create a very pleasing visual effect and design on their own through the selection of different styles and types of track light cannisters.

Decorative lighting is usually provided by chandeliers as the centerpiece in a dining area, foyer, family room and other areas. Its size will depend on the ceiling height and wall width of that particular area. In a dining area the chandelier should be approximately 12" smaller than the width of the table. The bottom of the chandelier should be about 30" to 36" above the table for proper lighting, and placement out of the direct line of vision when seated.

The diameter and length of a chandelier for an open foyer will depend on the height and width of the room. When centered in the area the bottom of the chandelier should not be less than 7' from the floor. Ask the lighting supplier for assistance when selecting decorative chandeliers.

Bath and vanity areas: Broadway lighting is one of the many ways to beautify your bath or vanity. They can be mounted individually, horizontally, vertically or joined together around a mirror as seen in dressing rooms. Broadway lighting strips have knockouts at the ends so they can be connected in a continuous run, and have as few as three, or as many as twelve lights. This lighting system uses more wattage, and creates more heat as the number of lights increases. It is therefore the least energy efficient. Overhead florescent strips are more economical, but require a valance or cover to hide their appearance.

Saving energy dollars: Although lighting accounts for only 15 to 25% of the electricity used in a home, this energy cost can be reduced substantially. One way is by using florescent lighting which gives more light per watt of electricity than incandescent bulbs. This way you get the same amount of light but use up to 65% less energy. A ceiling fixture that uses two 60 watt bulbs can be replaced with 2 - 20 watt florescent tubes. You save about 80 watts and get approximately the same amount of light. To further illustrate this point:

Incandescent	VS	Florescent
4 - 40 watt light bulbs	=	2 - 20 watt florescent tubes
2 - 60 watt light bulbs	=	2 - 20 watt florescent tubes
1 - 1000 watt light bulbs	=	2 - 20 watt florescent tubes
3 - 60 watt light bulbs	=	4 - 20 watt florescent tubes
2 - 75 watt light bulbs	=	4 - 20 watt florescent tubes

In addition to saving energy, fluorescent bulbs are excellent where a high level of evenly distributed light is required in areas such as the kitchen, hobby room, workshop, laundry and bathroom.

Other considerations for energy conservation include:

- Turn off lights when they are not being used.
- Put lights only where you need them.
- Use dimmers to reduce the amount of electricity used, and extend the life of bulbs.
- Finish your house with light colored, flat latex paint on the walls and ceilings.
- Keep all bulbs dust free. Dust and dirt can reduce the efficiency of your fixtures by up to 25%.
- Replace the old Incandescent 60 watt bulbs with 40 watt energy efficient florescent bulbs. The initial replacement cost is expensive, but they will pay for themselves within the first 6 months.

◆ SECURITY AND LOW VOLTAGE LIGHTING:

Safety and security are the main reasons for having good lighting outside your house. Good exterior lighting in dark areas around the house will prevent accidents and possible burglaries. Police recommend outdoor lighting and lighting with motion sensors as important security measures. When combined with interior lights on automatic timers preset to turn selected light fixtures on and off at pre-determined times in the evening, security is enhanced by giving the appearance of someone being at home.

The use of low voltage lighting in conjunction with exterior flood or walkway lights will add to the appearance and security of the house, and save energy dollars. Low voltage outdoor lighting provides safety for guests to see where they are walking, and for security purposes to see who is coming to the door at night. These lighting units can be on ground level, or set on various lengths of poles with a main, light-sensitive sensor mounted on or by the house to turn the lighting on at dusk and off at dawn. Many of the low voltage systems out in the market use 12 volt light bulbs, similar to the ones used in automobile brake lights, which cost less to replace than flood lights or wall mounted walkway lighting. These systems can be used as floodlights or area lighting to shine on a shrub or the house creating an impressive night time display of shape, color, and shadow around the entrance or walkway. Remember to consider the comfort of your neighbors when directing the lights. During the warmer weather, cost-effective outdoor lighting can extend the amount of time spent outdoors by lighting the grounds for barbecues, or just relaxing with the family.

◆ FIREPLACE SUPPLIER

Although the fireplace seems secondary to the construction of the house, it has to be one of the first items decided upon because it must be installed when the framer is on-site. Once the rough opening for the fireplace has been framed, the fireplace supplier will install the unit by bolting it to the floor or platform provided. The installers will measure and set the face of the fireplace flush with what will be the finished face of the drywall. The selected fireplace finish may be flush or protruding from the face of the fireplace, depending on preference, but make sure the fireplace installer has been informed of these distances before he installs the unit.

With energy efficiency being given such a high priority there are very few masonry fireplaces being installed in new construction. Selections are most likely a gas or wood burning, premanufactured, "∅" clearance metal fireplace. There are many options available with these types of fireplaces, however the most common for a wood burning unit is a gas log lighter. The plumber installs the gas unit during his rough-in so make sure the unit chosen has a log lighter capability, and the plumber has included the gas supply line and installation in his estimate. Gas fireplaces already come with the gas insert, and need only be connected by the plumber to the gas line.

Most fireplaces are considered an extravagance when energy efficiency is considered, as most fireplaces are not energy efficient at all. In fact, they can remove more heat up the flue than they put into the room. They require combustion air to burn, and in doing so draw the interior air from the house up and out the flue. This air loss must be replaced as quickly as it is removed for proper combustion to continue. If the fireplace selected does not have its own fresh air supply system, the air will have to be drawn in through cracks around the doors and windows, clothes dryer vents, bathroom or kitchen exhaust vents, and leaks between the house walls and foundations. The air required by the fireplace is usually warmed up to room temperatures before it reaches the unit, and many people consider throwing this warm air up the chimney a waste of heat and money. In addition, when the fireplace is on, the house heating system must work overtime to heat the air being drawn into the house from the outside.

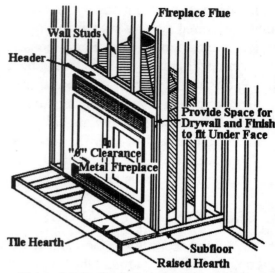

Figure 26-47. Illustration showing the typical framing for a "Ø" clearance fireplace.

Many houses are being constructed so air tight that a standard built-in fireplace would not be able to provide sufficient draw-air for the fire to continue burning. This results in a fire that smokes rather than burns, and sends smoke into the room rather than up the flue. To rectify this problem, fireplace manufacturers now may provide the fireplace with its own fresh air intake, and recommend the installation of air tight doors to the face of the fireplace which eliminates the removal of warm air from the house during operation. They become truly energy efficient when circulating fans are installed to take the heat from the super-heated metal walls and throat of the fireplace, and blow it into the room rather than send it up the chimney chase and flue.

 Note: Contact the fireplace installer for the diameter of the fireplace flue. Providing this information to the heating contractor as soon as you order the fireplace, will allow sufficient time for the heating contractor to cut the correct size of opening into the sheet metal chase cap to properly accept the fireplace flue. It is the responsibility of the fireplace installer to connect and seal the fireplace chimney collar after the metal cap has been installed.

Gas or wood burning fireplaces that are installed close to the center of the house or room require a vertical, fire-rated chimney flue to the roof for exhaust gases. If you have chosen a gas fireplace, and it is located on an outside wall, or within 6 feet of an outside wall a direct vent fireplace unit can be considered.

The hearth of a wood burning fireplace is simply a fire precaution against flying sparks, and may be set even with the floor, or raised above the floor level. Gas fireplaces do not require a hearth, but they are usually installed for the cosmetic effect of a real fireplace. The front fireplace hearth should extend from 16 to 18 inches out from the finished face of the fireplace, and is used as a display area or a comfortable place to sit on chilly winter nights. Confirm with the city building inspector the required hearth extension for your area as local fire regulations may vary this distance.

When planning a fireplace, locate it away from the

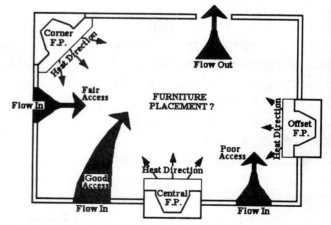

Figure 26-48. Illustration showing fireplace locations.

flow of normal traffic. Placing it on an inside wall will not block an outside view; outside wall placements allow cold air to be conducted into a room. (see Figure 26-48). It is a good idea to have at least 42" of wall length on either side of the hearth to allow comfortable furniture grouping. If the fireplace location is between or close to doorways or thoroughfares there tends to be more problems with furniture placement.

Wood Stoves:

People have become so energy conscious in the last 10 years they have forced the fireplace manufacturers to convert the old and traditional stove into a new technology, cast iron stove. These stoves have become visually more appealing with new heat-resistant, glass doors that allow a view of the fire. They are more energy efficient than their former models, and are capable of burning coal, wood chips, gas, or logs. There are many types and sizes of stoves on the market these days, and selection ultimately depends on personal heating requirements and tastes.

If information on a stove heater is required, contact a well established supplier with a good selection of stove units. They will be more than willing to show and explain the workings of different models that will service various needs.

♦ INSULATION CONTRACTOR

Although the insulator and vapor barrier installers are usually the same company, it will be more useful and easier to understand their uniqueness if discussed separately. Think of insulation as a cover or blanket. In order to keep the house comfortable and warm, it is necessary to insulate all perimeter areas to prevent heat escaping and cold air entering. Every area or level of the house should be insulated with different amounts or thicknesses of insulation because different areas allow heat loss quicker than others.

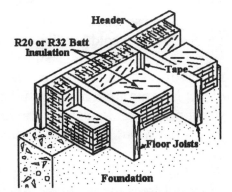

Figure 26-49. Drawing showing insulation requirements between the joists.

Basement:

The greatest outside area of a basement is generally below ground level. Since the earth retains a good percentage of its heat year around, it is necessary to apply insulation to that level of foundation that is effected by the winter cold. Each area in North America has a designated frost level, which is a depth below the surface that is effected by cold, and has a certain amount of winter freezing. Warm climates might have little or no frost, while other areas further north will have deeper frost levels. Basement foundations effected by frost levels must be insulated to stop the cold from radiating through the foundation and perimeter floor joist cavities into the basement and eventually throughout the house.(see Figure 26-49).

Building codes require insulation only to the level of the frost, however this is not practical if you plan on any basement development. Most builders recommend a full height, framed, frost wall around the interior perimeter of the basement foundation. The framer installs the frost wall which is usually a 2"x 4" framed stud wall, spaced approximately 24" on center, and checked for a vertical level with sufficient space between the foundation wall and the face of the frost wall to fit the R 12 or R 20 fiberglass insulation. (see Figure 26-50). At this time the insulator will insulate between the floor joists

Fugure 26-50. Illustration showing insulation requirements for cantilevers and frost walls.

around the perimeter with R 20 or R 32 batt insulation (see Figure 26-49), and at all cantilevers with R 40 insulation. (see Figure 26-50).

   Note: While the framer is on-site constructing the frost walls have him add, brace, replace and/or support any main floor wall studs or floor joists that have dried or twisted in order to meet building code requirements. The framer will use a crow bar or slug hammer to remove the bad wall studs, and hammer the exposed nails into the floor or ceiling plates. When installing the new studs the framer will toe-nail the studs into the plates; any replaced joists must be attached with metal hangers to pass inspection.(see Figure 26-51).

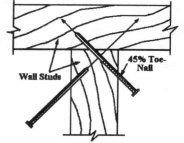

   Note: Make sure the framer provides a perimeter strip of acoustical sealant and vapor barrier under the bottom plate of the frost wall, or on the basement floor before he lifts and nails the wall in place.

Figure 26-51. Toe-nailing detail.

Why Caulk?

If there are any air spaces or holes in the floor joists or headers that have not been closed or sealed with caulking by the other subcontractors, have the insulator fill these gaps with insulation and seal them with acoustical sealant. This will eliminate any rodents from entering, or winter winds from blowing into the joist spaces and potentially freezing the water lines.

In our present house, I thought that I had accounted for everything, and double checked all the problem areas. One very cold evening our first winter in the house, my wife and I were sitting in the kitchen area enjoying a coffee when I spotted a movement under our refrigerator. At first I thought it was my imagination until I saw our 17 year old cat streak across the floor, and turn around showing us a brown, string-like object hanging from his mouth. I did not think that cat could still move so fast! On closer inspection the string had fur on it, and was moving. After several attempts to pry his jaws open, the cat reluctantly give up his prize, but the mouse by this time had expired so we thanked the cat, and put the mouse into the garbage.

We assumed the mouse had gotten into the house through an open door from the garage, and planned to set out traps the next day. About 3:30 that morning we heard a banging noise, and sounds of the cat running around the house. After what had happened earlier that evening, we knew what was going on. The cat was literally throwing a dead mouse into the air with his paws, and banging it into the walls. When this same scenario repeated itself the next two mornings, I knew that mice were getting in around a pipe or duct going to the outside that someone must have missed during insulating and caulking.

The next morning I had a handyman walk around the house and basement looking for an access hole, but he could not find one from the outside. I suggested examining the area underneath our deck, because that is where the built-in cook top vent and basement fan exhaust vents are located. The only way to access those areas is from the basement, so the handyman removed some joist insulation from several areas, and found mouse droppings by one vent unit. Reaching in he found that the furnace installer had forgotten to seal around one of the exhaust vents, so I had him do a quick patch job by packing insulation around the vent to temporarily seal the access. I had to wait to seal the exterior with caulking until the weather was warmer, and someone could get underneath the deck. Even after years of being in the design and construction industry, one can still miss something that appears to be minor, but when you think about it, we lost several nights sleep, and most likely a fair amount of heat through that hole.

Main Floor:

Insulation of the main floor is very important in maintaining all areas of the house at a comfortable temperature year round. Because this area is totally exposed to the elements, it is necessary to have all perimeter walls, corners, windows, doors and exterior drill holes insulated to eliminate as much of the cold air infiltration as possible.

When the electrician and plumber have finished their rough-in, and after inspecting their work you may be inclined to think that about 50 woodpeckers helped them, but got out of control. There are holes drilled in the exterior and interior wall studs, top and bottom wall plates, exterior plywood sheathing, and through the subfloor of the house. These areas totalled are potentially equal in size to an uninsulated 4" wide by 3" high hole in a wall of your house year round.

All the inaccessible corners of the walls in the house should have been insulated by the framer during the framing stage. The insulators will then seal all potential air leaks in the house with insulation and acoustical sealant to prevent blowing air entering the walls from the attic and basement. This should include the sealing of all holes around electrical wires on interior and exterior stud walls, as well as all holes left by the plumber and electrician in the upper and lower wall plates Air spaces that are too large to be sealed with acoustical caulking, such as oversized holes around the plumbing vent stacks and basement drain lines, should be sealed with a thick rubber gasket supplied and installed by the plumber. If this is not available the insulator can pack the hole with insulation, cut a left-over insulation stop to fit tightly around the pipe, and then caulk and staple it to the plates. Make sure that your estimate from the insulator includes this service. (see Figure 26-52).

Once this has been completed the insulator will place friction fit, fiberglass batt insulation between all the studs, around all window and door frames, and in all roof areas not accessible after the vapor barrier and drywall are in place. These hard to reach, narrow roof areas are inaccessible to the insulator once

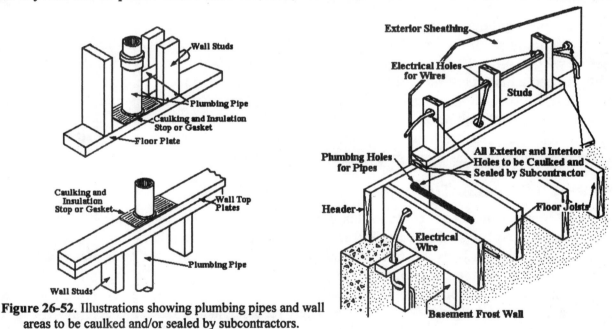

Figure 26-52. Illustrations showing plumbing pipes and wall areas to be caulked and/or sealed by subcontractors.

drywalled as they will not allow him access to blow in standard, loose fill insulation. These narrow spaces are usually a result of cathedral or open beam ceiling requirements where the contractor is using narrow trusses, or 2" x 12" roof/ceiling joists. Discuss these areas with the insulation installer so they can install the proper amount of batt or fiber insulation (usually a minimum of R 32 insulation) prior to application of the vapor barrier. Remember, these areas will require a minimum 3" air space between the insulation and roof sheathing for air circulation. If this space is not provided for proper air circulation, it will cause winter freezing followed by warm weather thawing resulting in major water damage to the adjacent structures. Proper soffit and roof ventilation must also be provided in the joist spaces of these areas for the best air circulation.

As discussed earlier, certain interior walls such as plumbing walls, and walls between bedrooms and other areas of the house may require sound proofing using batt insulation. The thickness of insulation will depend on the wall. Most bathroom walls are from 5 1/2" to 7 1/2" thick so an R 20 batt should be sufficient. Make the insulators place the insulation so that both sides of the pipe will be covered with batts. Most other walls are 3 1/2" thick and would therefore require R 12 batt insulation as sound proofing. (see Figure 26-53).

A well insulated house will keep the rooms cool in the summer, and warm during the winter, but it is necessary to inspect the job to see that it is being done correctly. A contractor who spends most of his time running around trying to keep everybody happy, or has too many other houses on the go at the same time will not be able to provide proper supervision of jobs such as this. As a result omissions and discrepancies show up after occupancy. Our personal experience

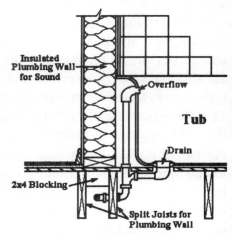

Figure 26-53. Illustration showing insulated plumbing walls.

relates to drafts that seemed to be located in the family, kitchen, and breakfast nook areas. After checking the furnace vents to see if they were blowing adequate air, the windows for drafts, and the walls to see if they felt cold I decided to do the old match test. This simple test involves lighting a match, blowing it out, and seeing where the smoke goes. After lighting half a package of matches, all the smoke seemed to be going rather quickly toward the ceiling. Time to check the attic insulation. Guess what, there was none!!! No wonder it was drafty as all the heat was being sucked out though the ceiling. I was very glad that I caught it before we got minus 10 and 20 degree winter weather, but I was very angry that the builder had been so negligent about his supervision. I contacted my own insulator, and had him install the required R 40 (10 1/2") loose fill insulation throughout the attic area that same day. So remember, do your own inspections even if a general contractor is building your house.

Cleaning note: It is the responsibility of the insulation installer to clean and remove all the leftover insulation containers from the house, and take them to a disposal site.

VAPOR BARRIER INSTALLATION:

It is the vapor barrier that stops the interior humidity caused by normal use of household appliances such as the dishwasher, clothes washer, baths and showers as well as body humidity from passing into the building structure. If the vapor barrier is not installed, or installed improperly, the moisture contained in the house would transfer into the cavity of the wood structure where there is sufficient wind and cold to cause condensation and freezing on the studs and walls. This moisture would constantly freeze and thaw, and in time probably cause rotting and decay of the structure reducing its life span considerably. The most important feature of a vapor barrier is its continuity as it acts as an air barrier between the warm interior and the cold blowing winds on the exterior of the house. The vapor barrier must be strong enough to withstand the wind pressures which can occasionally reach very high velocities within the roof and walls. A 6 mil polyethylene vapor barrier is most widely used as it is thick enough to stop the radiating vapor pressure, and strong enough to resist the wind pressures. (see Figure 26-56).

The electrician will have supplied and installed poly vapor hats around electrical outlets on all main and basement outside walls and cold ceilings to eliminate those air leaks; the framer will have provided all the necessary vapor blankets at the top and bottom plates to ensure a continuous vapor barrier throughout the house. When installing the vapor barrier over the vapor hats and vapor blankets a thick layer of acoustical caulking is applied prior to placing the main vapor barrier on top. This securely seals the house vapor barrier to the hats and blankets. Where the electrician's wires enter the poly vapor hats, the outside is sealed with tape or caulking.

The 6 mill polyethylene film is available in large, room-height sheets allowing continuous application with a limited number of joints as reducing the number of joints reduces the chance of air infiltration into the house. Where the sheet ends and a joint is necessary, the new and finished sheets of poly vapor barrier must be sealed with acoustical caulking at a stud, and stapled at the end and start of the lapped sheets. The sheets should overlap by at least one adjacent vertical stud, and be stapled on top of the caulking to seal the joints and eliminate air infiltration.

The ceiling vapor barrier should also overlap the wall vapor barrier, and be sealed with acoustical caulking at the intersection of the wall and the ceiling. In addition, the vapor barrier is stapled over the roof truss caulking to maintain continuity. Although the interior partitions are framed before the vapor barriers are installed, this continuity problem is resolved by the framer covering the top and ends of the interior walls with a strip of vapor barrier that is wide enough to allow sufficient overlapping with the ceiling and wall vapor barriers. The framer walks on the tops of the interior partitions when installing the roof trusses so in order to avoid damaging the vapor barrier strips, and also to

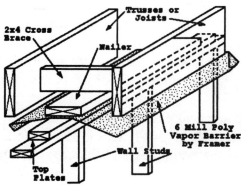

Figure 26-54. Vapor strip drawing.

provide better footing for the framer's crew, the vapor strips are installed between the two top plates. (see Figure 26-54).

The single, very large, consistent error made by builders is not stopping the air from blowing in around door and window frames. Most frames are constructed so that when installed by the framer they will protrude into the room the thickness of the drywall. This gives the finisher a base on which to nail the door and window trim, but leaves an unprotected crack around the frame. Vapor barriers are usually overlapped and stapled, but not caulked to the door and window frames by the installer. When the drywall board is later applied, the vapor barrier is inevitably torn or pulled away from the frames. This unprotected crack can prove costly, and be eliminated if the drywaller's estimate, which usually includes the installation of the vapor barrier, states that the vapor barrier installer "will return to apply a bead of caulking in the gap between the drywall and door/window frames, and add a strip of cellophane sealing tape to the frame and the drywall". This will all be covered when the finisher nails his trim in place. (see Figure 26-55).

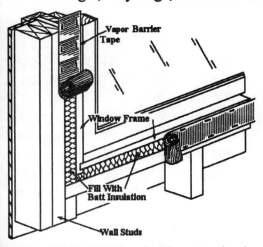

Figure 26-55. Drawing showing vapor barrier around all door and window openings.

It is also important to provide a vapor barrier protection to the insulation installed between the floor joists. This area is very difficult to protect because the vapor barrier must be cut to fit between the joists, and the installer must sufficiently caulk the joists and subfloor so that when the poly is stapled it will provide an air tight seal all around. When floor joists in two story homes require a vapor barrier the caulking around any batt insulation becomes very difficult, therefore a hard surfaced, rigid insulation is better suited. A double thickness of 2 inch rigid insulation fitted tightly between the joists with acoustical caulking around the perimeter allows the installer to press and better secure the vapor barrier in these tight areas. Extra care should taken at the exterior header area to prevent tearing of the vapor barrier. A double layer of building paper or a tyvek paper wrap should be considered in these areas to help minimize air infiltration.

The weakest point in any vapor barrier of a house is the attic or hatch access. The plan may require only one inside the house, but it should be properly weather stripped, and preferably located where it will cause the least air loss, eg., in an attached garage, or on an outside gable wall not visible from the street for security reasons.

Cleaning Note: It is the responsibility of the vapor barrier installer to clean and remove all the leftover polyethylene from the house, and take it to a disposal site.

Note: The exterior finish on the house should be completed before the drywall is installed. The pounding of nails when applying the building paper and exterior finish will loosen the nails or screws from the drywall. This is an example of where the timing of the subtrades and installers is very important to the quality of the finished product. You may proceed to pages 139 to 143 on exterior finishes then return to drywalling, or just continue keeping that chapter in mind. This added side-step in the construction process will be accounted for in its proper sequence in chapter 31 on scheduling suppliers and subtrades.

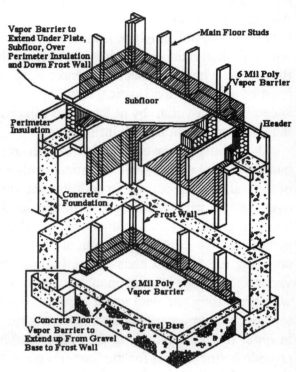

Figure 26-56. Illustration showing vapor barrier location.

◆ DRYWALL CONTRACTOR:

With progression to installation of the drywall comes the realization that the house is finally moving to the finishing stage where the actual room sizes can be seen, and not just visualized on working drawings.

Drywall or "gypsum board", which is basically a gypsum filler sandwiched between two sheets of paper, is the most widely used wall finish due to its speed in application, low fire-rating, low cost, flexibility, variation in size, and consistent quality of finished surface. The edges on one face along the length of the board are slightly tapered so a joint filler can be used without causing a ridge. The thickness of the board application depends on the spacing of the wall studs, or the required fire-rating. Many contractors building the lower end or starter homes use 3/8 inch drywall for wall stud spacing of 16 inches on center to take advantage of the lesser cost. The 1/2 inch drywall board is more commonly used in construction because of it additional strength.

In some energy efficient homes where the wall studs are 24 inches on center, the minimum thickness used has been a 5/8 inch drywall board. Where the house has an attached garage, the common wall between the garage and the house must have a 1 hour fire-rating, therefore a fire-rated 5/8 inch gypsum board is used. In areas of high humidity and water, eg., tubs and showers, the drywaller installs a water-proof aqua board.

The drywaller measures the house from the blueprints, and calculates the longest length of board to be used with minimum cutting for each wall or ceiling area. The drywall is applied directly to the framing member with a minimum number of ringed nails or screws to limit potential nail-popping. If nailing is quoted on the estimate, ensure that the nails are driven into the drywall in pairs at intervals of about 12 inches along the studs, joists or trusses. This double nailing procedure will help to reduce nails from popping. Pricing for drywall screws should not cost more than the double nailing method. The main differences are the screws will fasten themselves more tightly to the wood structure eliminating popping almost entirely, and they can be fastened every 16 inches rather than the 12 inches required for nails. Drywall can be attached with ringed nails, screws, or a combination of the two. If you can afford the additional cost of labor, and apply glue in a continuous bead to the vapor barrier along the framing member, this will help seal the nails or screws against the vapor barrier thus eliminating any air infiltration, and making the house more energy efficient.

The drywaller usually installs the ceiling board first, placing the long dimension at right angles to the joists or roof trusses. The ceiling boards should be staggered so that the butted ends of the boards do not all attach to the same joist or truss. This produces a stronger roof structure when nailed or screwed, and eliminates the possibility of a continuous running ceiling crack. Drywall is most commonly applied to walls in the longest lengths possible, and set horizontally rather than vertically to reduce the amount of cutting and number of nails or screws required, and allow for staggering of the joints.

The drywall contractor is required to provide a warranty service call after one year at which time he will return to the house, and repair all the cracks and nail pops. The drywaller is responsible for only the repairs to the drywall, so the wall areas restored will be finished with drywall cement which will have to be painted at your expense. I suggest that this is the time to have the painter return to apply the second coat of paint to the house including a double application of paint in the areas of drywall repair. This will guarantee a more uniform paint surface throughout the house.

Cleaning Note: It is the responsibility of the drywall installers to remove all the leftover drywall board from the house, and take it to a disposal site.

Taper:

The taper removes all loose paper from drywall board, and cleans all the joints of mud or loose dirt which could cause discoloration of the joint cement, or prevent the cement from adhering to the drywall. For the first step of the process all joints and nail or screw heads are covered with joint cement and allowed to dry. All external corners are protected with corrosion-resistant or plastic corner beads. For all interior corners the tape is folded in half and applied with joint cement; for all horizontal and vertical joints the tape is applied full width with joint cement. If the interior temperature is less than 20 degrees centigrade or 68 degrees Fahrenheit, the furnaces must be connected and running to provide sufficient heat to dry the joint cement.

Note: If at this time during construction the gas line to the furnaces has not been installed, supply the taper with alternate heat to dry the drywall joint cement. Propane blowers are most commonly used, and the propane cylinders and blowers can be rented weekly or monthly. Propane is a moist heat, and it will therefore require more time to dry the joint cement. If building during the fall and winter, this type of alternate heat can be extremely expensive, therefore try to schedule the installation of the meter and connection of the furnaces as soon as possible.

After the first layer has dried overnight the taper will apply the second layer of joint cement to the corners, horizontal and vertical joints, and the nail or screw heads. The edges of the joint cement strips are feathered so that a bulge at the joint is eliminated. (see Figure 26-57). Once again sufficient heat must be available for the cement to dry properly.

A third layer of cement will then be applied to all corners, joints, and if required, any nail or screw heads. This third and final layer of joint cement is applied, and tapered wider and smoother than the first two layers so that he will have to sand only slightly to achieve a smooth, flat look to the cemented areas of the drywall. The final application must not have any bumps, bulges, or rough edges in preparation for painting.

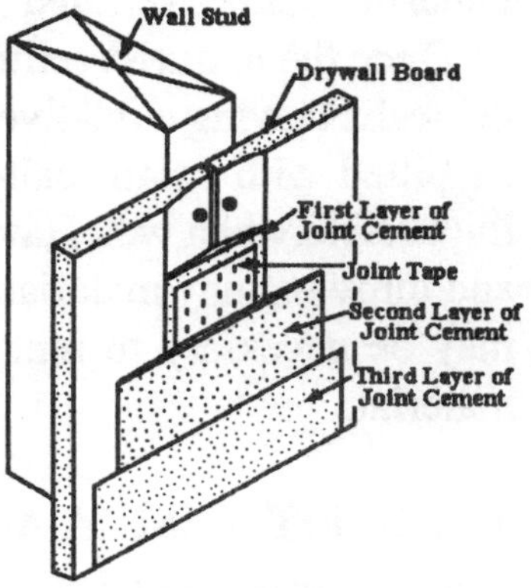

Figure 26-57. Typical board taping application.

Many custom home builders request the drywall corners be finished with a rounded plastic corner bead. The application of cement is similar, except more care must be taken when adding the cement to the corners. The cement must be feathered out further to give the rounded corner a smoother appearance. Rounded corners are more expensive, but they produce a cleaner look and added resale value.

Cleaning Note: It is the responsibility of the taper to clean the floor with a scraper to remove all the dropped cement, and remove all the leftover tape from the house, put it into boxes, and throw it on the designated scrap pile for removal.

Textured Ceilings:

Note: Before the ceiling is textured a primer coat of paint must be applied to the walls. Texture when dry is very white, and even a white paint is not as white as a finished textured ceiling. This primer is usually applied very quickly with a paint sprayer or roller, and if the ceiling was already textured the paint from the sprayer or roller would discolor the texture, and result in expensive re-application to all the ceiling areas. This additional side step in the construction will be accounted for in its proper sequence in chapter 31 on scheduling suppliers and subtrades.

A textured ceiling is commonly used throughout the house because of its speedy application, cost effectiveness, and the consistent high quality of the finished surface. A textured ceiling also saves the home builder the additional costs for taping and sanding the ceiling to a smooth and finished surface ready for painting. To prepare the ceiling drywall board to accept the texture, the joints, nails and/or screws are covered twice with drywall cement and smoothed with a hand trowel. It is usually only necessary to lightly sand the ceiling at corners and questionable surfaces to eliminate any rough or bumpy areas which might show through the texture. Once the rough sanding has been completed, the walls are draped with sheets of polyethylene so that when the texture is applied to the ceiling any overspray will stick to the poly rather than the painted walls. The ceiling is then machine sprayed with a thick, even compound of drywall cement which once dried is called texture.

More custom home builders are leaning toward a more expensive and fashionable style of ceiling texture called "California knock down". To prepare the ceiling for this style of texture, the taper provides a smoother, sanded ceiling at corners and all surfaces to allow the drywall to accept a colored coat of paint. If not properly sanded the drywall joints will show through this texture style. When the painters apply the primer coat of paint to the walls, the ceiling is also painted. Usually a soft pastel color is chosen for the ceiling, so when the texture is sprayed in a more random pattern, the paint will show through the texture. The sprayed cement is then lightly trowelled to flatten the finished appearance. The stylish effect of the completed ceiling is very hard to explain, other than if a blue pastel color is selected as the background paint it will appear like white, billowing clouds passing over a blue sky background. In my opinion it looks fantastic, but is overrated for the price.

Once the textured ceilings are completely covered, the tape and polyethylene covers are removed from the walls leaving a relatively clean house except for the floor. Because of the spray technique the floor will be pitted with small, ball-bearing sized, white particles which are very hard to remove when dry, and impossible when wet. Have included in the price estimate scraping the texture particles from the subfloor, and throwing any materials outside on the trash pile. In order to get all the texture particles off the floor it may be necessary to rent a commercial cannister vacuum cleaner designed to pick up this small, fine material.

◆ SOFFIT, FASCIA AND EAVES TROUGH INSTALLER:

The soffits, fascia, and eaves trough of the house can be started after the roof shingles or shakes have been installed, and before the exterior finish is applied. The downspouts for the eaves troughs should be installed after the exterior finish is completed because they must be attached with screws and strapping to the exterior walls.

Many suppliers of soffits, fascia, and eaves troughs also install the siding which reduces the running around to collect estimates, and makes the installation scheduling much easier. However, if stucco, cedar, brick, or any product other than siding has been selected, it will be necessary to have three estimates from suppliers that also do soffits, fascias, and eaves trough installations. These installers are not hard to find, but the scheduling for the different material installation by the different subcontractors must be coordinated correctly, so as not to cause any construction delays or installer callbacks.

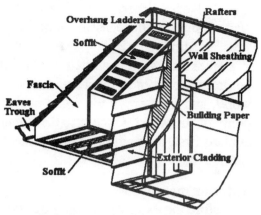

Figure 26-58. Typical gable end soffit and fascia projection drawing.

The fascia material comes in long flat rolls of prefinished sheet metal, and a wide variety of different colors. These prefinished sheets are bent by machine on the site, and molded into long lengths of 6" fascia which can be handled and usually installed by two men. Shorter lengths can be made if one person is installing the material, however, the fewer the number of joints the better the fascia looks, and the less the chance of water seepage. The fascia installation is as simple as taking the premanufactured length, sliding it up and under the finished roofing material, and nailing or stapling it to the wood rafter or truss projection header board. (see Figure 26-58).

The soffits come from the manufacturer in cardboard boxes cut to specific lengths with precut ventilating holes for quick and simple installation by the contractor. A starter strip of building paper must first be placed on the wall of the house at the level where the installer will staple the soffit sheet. The fascia manufactured on-site will have a ledge to hold the soffit section in place so that once the soffit is fitted on top of the fascia ledge they are both nailed together to the underside of the wood rafter or truss projection header board. (see Figure 26-59). Soffits must be placed under any protrusion or cantilever of bay or box windows in order to provide a finished and waterproof surface under the protrusion, and hold the insulation in place.

The eaves trough and fascia material are manufactured by the same company so that the color and product finishes will match. They both arrive on-site as long rolls of flat, prefinished sheet metal which are then cut to a specific width to fit into a portable shaper installed on the contractor's truck. The shaping machine can be adjusted to make several different eaves trough widths, and is capable of molding the sheet metal into single lengths of troughs that will fit almost any distance required.

The eaves troughs are secured in place with long aluminum spikes, and randomly nailed through the fascia into the wood rafter or truss projection header board. (see Figure 26-59).

Cleaning Note: It is the responsibility of the fascia, soffit and eaves trough installer to pick up and remove all sheet metal debris from the roof and perimeter of the house, and place that material on the trash pile.

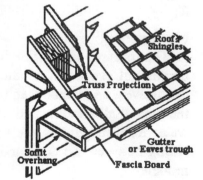

Figure 26-59. Typical eaves trough projection drawing.

◆ EXTERIOR CLADDING AND INSTALLATION:

As previously discussed in vapor barrier installation, page 136 of chapter 26, it is critical that exterior cladding requiring any nailing be completely installed prior to nailing or screwing the drywall. The hammering vibration on the exterior of the house will loosen the interior drywall nails causing them to pop, and require servicing which subsequently adds extra cost to the house builder.

Note: When reviewing the cladding installer's estimate, check for the statement that they will supply all necessary labor and materials to complete the job, including scaffolds, flashing, and caulking.

Before any work is started on the exterior walls, any wood surfaces that will remain exposed must first be protected from moisture with a sealer, and a finished coat of stain or paint. The painter will clean the wood, and then apply the wood sealer on both face and side surfaces for maximum penetration into the pours of the wood. Once dry, the finish coat of oil base stain or paint is applied. If the exterior cladding is applied first, the painter cannot protect the side surfaces of the trims, and may apply paint smudges to the siding.

Exterior membrane:

Before the exterior cladding can be attached to the walls, the studded wall cavity and exterior sheathing must be protected with a water resistant, but vapor permeable building paper, or tyvec wrap which is stapled to the exterior sheathing every 6 inches. Its main function is to stop any wind or rain that might pass through the exterior finish. It must be vapor permeable to allow the escape of any water vapor that may enter the stud cavity from the interior or exterior, and allow the house to breath through the cracks or building imperfections. The building paper or tyvec wrap should cover the entire face of the exterior wall with 4 inch laps for building paper, and 8 inch laps for tyvec wrap at all end joints. The areas of concern to protect on a house with double layers of this membrane are all cantilevers, corners, and around all window and door openings. For optimal protection insist that the membrane be installed on all wood surfaces (horizontal or vertical) to the bottom of the exterior cladding and within 8" of the finished grade. (see Figure 26-60).

Promoters of super-insulated homes recommend rigid insulation and additional strapping be attached to the exterior sheathing of the house. The rigid insulation is covered with a wind resistant, but vapor permeable house wrap, and the exterior cladding is then attached to the wood strapping which is spaced approximately 24" apart in a vertical pattern. This added insulation is best suited for the north where the heating costs are very expensive, and winter temperatures reach well below the freezing point for many months. Weighing the additional costs against the payback term might not deem this step cost effective for many regions.

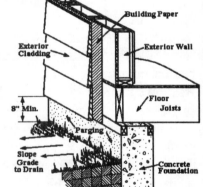

Figure 26-60. Typical grade to cladding height drawing.

Flashing:

The protection of the house against blowing snow and rain is paramount especially in those areas that are not designed to protect themselves, i.e., window and door openings. The installer will provide a water-proof strip or drip cap over the tops of all window and door openings prior to the installation of the exterior cladding. This will stop water from infiltrating the trim through the exterior membrane and sheathing, and possibly ending up sitting between the door and window jams to damage the structural walls and drywall with dry rot or mildew.

Reminder: Have the framer install the 2"x10" wall header to the future deck attachment. This way the exterior cladding installers can also add the water-proof flashing to the top of the header.

Exterior claddings:

The exterior finish will significantly affect the streetscape impression and re-sale value of your house. Select with care the color, style, type, and quality of cladding as this choice will directly affect the amount of leisure time you and your family are able to spend together, as opposed to putting in time maintaining the cladding and exterior finishes. The most common and economical types of cladding are: metal, vinyl and hardboard sidings, stucco, and masonry [brick or stone].

All exterior cladding or sheathing that can be effected by moisture must be kept a minimum distance of 8 inches from the ground or finished grade of the house. Check with the local building inspector about the area's height requirement. (see Figure 26-61).

It is also very important to make a final inspection of the house before the exterior cladding installers leave the site. With working blueprints in hand, review the exterior electrical plug and light locations to make sure they have not been covered up with cladding. The electricians when installing the fixtures and plugs might miss them, and assume they have been covered for a good reason, decide to find them by punching poles in the cladding, or say that they will not return until the cladding installers return to find the buried electrical boxes. Either way, it becomes very inconvenient, and a final inspection can eliminate many potential problems, and ensure a quality finished product.

Metal and Vinyl sidings are the most commonly used cladding, and are considered a maintenance free product because the finish is heat baked at the factory. Manufactured for simple installation, the boards interlock together, and are nailed or stapled from the top of the siding board into the exterior sheathing of the house. The manufacturer sends the siding out prepackaged in virtually every color and tone of the rainbow with names such as `double 4´ and `double 5´.

When these materials are installed in long lengths on continuous walls with a direct south or west exposure, the sun will cause substantial expansion and contraction sometimes causing the siding to warp and not return to its original shape. To limit this warping lightly nail or staple the siding which allows for more movement in these areas vs. hammering it tightly to the wall as in the other areas of the house. These materials are very easy to replace, and the potential warping should not be a major concern when receiving estimates.

For horizontal application of the siding, the first board will be fastened at a corner 6 to 8 inches above the finished grade of the house, or for a high basement the starter strip will be fastened about 3 to 4 inches below the exterior sheathing level. The same rules apply to vertical installation of the siding.

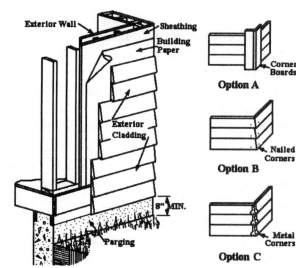

Figure 26-61. Illustration showing different corner siding applications.

When installing the siding, special trim pieces are manufactured to cover the end cuts and rough corners in such places as the perimeter of windows and doors, the intersections of siding and soffits at the gables, and all interior and exterior corners. The trim pieces around the windows and doors are installed before the siding; the starting point is at a corner with the application of the first board and corner trim together.

Cleaning Note: It is the responsibility of the siding installer to pick up and remove all debris from the roof and perimeter of the house, and place that material on the trash pile.

Stucco cladding is being more widely used as a feature material on the current California homes with finishing styles of arches and built outs. Homes of the 1950's mixed crushed glass, washed rocks, stones and even marbles into the usually bland, grey base to give it some color. Now with the many different additives and pigments the choice is from almost any color with lighter or darker tones of those same colors. The stucco installer must mix the right portion of cement, sand, and lime with the correct amount of pigment to match the brochure sample, but once he knows the specific proportions he will be able to produce it in sufficient quantities to finish the entire house. The cheapest and most standard stucco is white, and there is an extra charge for a color additive mixture.

The application of the water resistant, vapor permeable building paper or tyvec wrap is the same as for the other exterior claddings. Stucco however is not able to directly adhere or stick to paper, therefore the stucco installer must next apply a galvanized wire stucco mesh to the wall. The mesh is stretched and nailed horizontally into the wall sheathing over the building paper, and all wall joints in the mesh are nailed and lapped over each other by at least 4 inches. The corners, door, and window perimeters are all reinforced with a special, tighter knit metal mesh to allow the base coat to adhere better in areas where potential cracking and water penetration might occur.

The first coat, called the 'scratch coat', is forced into the reinforcing mesh with a hand trowel so that when completed the wire mesh will be completely covered. Before it dries completely the stucco surface is scored with a wire or hard bristle brush to provide a bonding key for the second coat. The installers will allow the scratch coat to set for about 48 hours, or until dry, before applying the second, or 'brown coat'.

Before this second coat is applied, the scratch coat must be dampened with water to guarantee a proper bond between the two. It is usually pre-mixed with the color pigment, and is thickly applied with a hand trowel to completely cover the scratch coat. In addition, it is usually more fluid in consistency, so when added to the scratch coat will allow a slow cure process to occur over the next few days. This will provide a better and more consistent adhesion to the scratch coat.

The third and final coat is called the finished coat. As before, the second coat must be dampened with water to also ensure a proper bond. The finished coat is then applied with a smaller hand trowel to give the stucco the desired textured finish, and the house its California-troweled appearance.

In warm, dry weather the stucco should be kept damp with a light sprinkling of water for several days to ensure proper curing. During cold weather it will be necessary to provide propane heat for the stucco so the mortar will be able to cure and set properly. This will require a consistent, warm temperature for a period of 2 to 3 days. If cracking or spider web striations appear, the stucco most likely was not allowed to cure properly, eg., too much or too little water in the mix, too cool in the evenings, or too many dry, windy or sunny days during the curing. If this happens, recall the stucco installer for servicing the next year during better weather. Having this service included in his estimate may save a few heated arguments, and several hundreds of dollars in extra charges.

For a two story house, the stucco workers will provide scaffolds to work from as this saves time by eliminating the climb up and down the ladder to collect and mix their materials. (see Figure 26-64).

Cleaning Note: It is the responsibility of the stucco installer to pick up and remove all debris from the roof and perimeter of the house, and place that material on your trash pile.

Masonry cladding is the most expensive exterior finish, but requires zero maintenance. The application of the water resistant, vapor permeable building paper or tyvec wrap is the same as for other exterior claddings.

Because of its weight the brick or stone veneer requires the foundation to carry a substantial amount of this weight. If the veneer height is one story or less, the builder can use a steel angle iron bolted directly into the concrete foundation as a structural supporting ledge. (see Figure 26-63). If the veneer exceeds the single story height it is suggested that the foundation be cribbed and poured to incorporate a structural supporting ledge wide enough to allow the width of the veneer plus 1/2 to 1 inch air space between the brick or stone and the

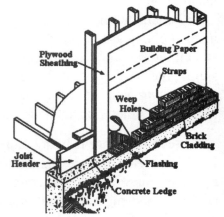

Figure 26-62. Illustration showing a concrete brick support ledge.

building paper. A galvanized strip of flashing placed on the supporting ledge should cover the full surface of the ledge, and extend horizontally beyond the ledge to act at a drip cap, and vertically 6 inches up the exterior wall behind the building paper. (see Figure 26-62).

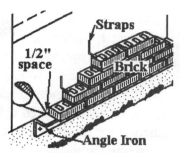

Figure 26-63. Angle iron brick ledge option.

Regardless of whether the angle iron or the foundation ledge is used, the brick must be fastened to the exterior structural wall of the house with L-shaped, galvanized metal tie straps. The straps are nailed vertically every 24 inches to the wall studs of the house, and the angled section embedded into the mortared section of the brick layers every second stud width, or 32 inches depending on the stud spacing. These straps tied to the structure of the house assist in keeping the brick or stone veneer vertical to the walls, and prevent the veneer wall from bowing out and collapsing.

Because mortar and brick are porous, water is allowed to enter the air space provided between the veneer and the house wall. This water must not be allowed to accumulate to cause potential water and frost damage to the house structure. The brick or stone mason provides weep holes to ventilate the cavity, and allow any water to drain out from behind the veneer. This is accomplished by eliminating some mortar from the vertical joint every third to fifth brick, or at 32" intervals along the bottom course. (see Figure 26 -62). The outside mortar joints of the brick should be scored to a smooth finish to provide protection against the penetration of water and blowing snow.

If a brick veneer has been selected, it should be locally manufactured, or reclaimed from a building within your weather zone. This will ensure that the brick will withstand the climate changes for your building location. Many local manufacturers of brick have in their showroom a large selection of styles, shapes, and colors that will suite your visual needs. They can even reproduce the appearance and color of 50 year old reclaimed brick. Stone veneers as well should be selected from local materials and quarries that supply several contractors within your area.

Figure 26-64. Illustration of brick worker using wood planks, and ladder scaffolds.

For two stories of brick veneer it will be necessary for the brick layer to provide scaffolds as a safety ledge to work from. The scaffold is also used to store the bricks and mix the mortar to be placed on the face of the house without having to climb up and down to collect these materials. (see Figure 26-64).

During cold weather it will be necessary to provide heat to the brick face to allow the mortar to cure and set properly. The temperature must be kept constant and above 10 degrees centigrade or 40 degrees Fahrenheit by using propane and propane heaters, and covering the working area with a tarpaulin or cover. The heat required to keep this area at a constant temperature cannot be provided by the house heat as it would be too costly. As these scaffolds, tarpaulins, and heat are all considered extra costs, try to schedule construction so that all the exterior work can be accomplished during the warmer summer months.

Cleaning Note: It is the responsibility of the brick layer to pick up and remove all debris from the perimeter of the house, and place that material on the trash pile.

♦ **KITCHEN CABINET SUPPLIERS:**

Once the primer coat of paint and textured ceilings have been completed, the plywood overlay is installed over the subfloor by the finishing carpenter before the cabinets and vanities are installed. This will eliminate any additional cutting, planing, or filing of the plywood overlay in order to fit under the cabinet toe kicks, and odd corners and angles caused by the cabinet shape or design. This will also protect the cabinets from being scratched or gouged by the workers when laying and stapling the overlay. If the overlay is installed after the cabinets, and filler pieces are used, this could result in an uneven floor surface for the finished floor material, and the filler seams and staples could show through the lino.

Kitchen cabinets present an attractive appearance, and help to increase kitchen efficiency. A well designed and arranged cabinet system with special attention given to the location of the refrigerator, sink, dishwasher, and cooking center will reduce work, and save steps between the different work centers. Here, as in other areas of the house, storage cabinetry should be designed keeping in mind the items to be stored. The house must be measured, and the space carefully divided with drawers, shelves, and work centers correctly proportioned to satisfy the specific needs of each area.

There are three types of cabinetry that can be considered depending on style, budget, and design: cabinets built on the site by a carpenter, custom built units constructed and assembled in a local cabinet or millwork shop, and mass produced units available from factories that specialize in cabinets. Even when a large part of the cabinets are produced in a shop, the carpenter is responsible for the installation which is a job that requires skill and careful attention to detail. The final cost of your cabinets will depend on the materials selected, the shape, number and style of cabinets, and the quality of counter top selected.

In all cases, the house must be measured by the sales representative to determine the size and shape of cabinetry required to fit each area of the house. They will return to their office, and from working with your blueprints and their on-site measurements draw a detailed sketch of the cabinets for the kitchen, bathroom and laundry that will conform to their modular sizes and your design requirements. Location and size of the appliances must be known by the sales person in order to match cabinets with the appliances; each work center must be discussed in detail so that the proper size and type of drawers and cabinet doors can be sized and drawn with the correct swing. (see Figure 26-66). Most refrigerators and ranges protrude beyond the face of the cabinets. It is possible to have the cabinet and the appliance face flush if the appliance is located on an interior wall. This is made possible by having the framer reduce the walls stud width by 2 inches thus allowing those appliances to be set further back into the wall. Have the cabinet sales person

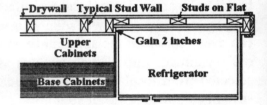

Figure 26-65. Drawing showing refrigerator set into typical wall.

design the kitchen layout to accommodate for this feature if you like the idea. (see Figure 26-65).

When the drawing is completed, sit down with the representative to discuss and verify in detail the location, type, swing, and size of every drawer, door, shelf, and appliance in each room. (see Figure 26-66). Sometimes if the sales person gets too busy, they will forget a feature that was discussed, and then assume that you wanted something else, or forget it completely, eg., a certain drawer size, door finish, or cabinet style. If they have not made note to call and ask questions it might be missed in the estimate. If not caught before ordering, there will be an additional cost when purchasing another cabinet to correct the mistake, or otherwise deciding to do without.

Always review in detail the following items twice with the salesperson, i.e., at the initial meeting to discuss the design, and after completion of the kitchen drawing with the final estimate.

- sizes of cabinets - width and height.   - locations of appliances.
- sizes of appliances.   - size of sink.
- door swing and hinge location.   - door knob style.
- number of shelves in each cabinet.   - name and style of cabinet face.
- style of pull-out for drawers.   - corner cabinet shelves or lazy susan.
- locations for cutlery, hand/dish towels, detergents, pots and pans, plates, drinking cups, containers, canned goods, mops and brooms, food staples, eg., flour, sugar or baked goods, etc.

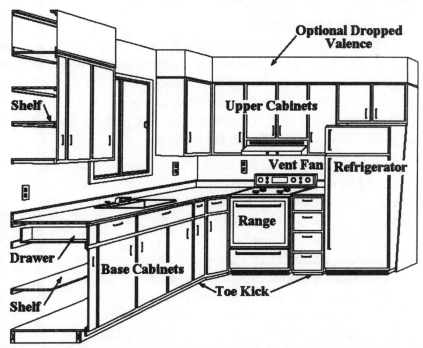

Figure 26-66. Schematic drawing of typical kitchen cabinets and appliances.

- refrigerator set into a wall in order to have it flush with the cabinets.
- distances between cabinets for circulation, especially around an island unit.
- dropped counter areas in the kitchen.
- location of the correct number of vanity sinks.
- supply and installation of the sunshine or island florescent ceiling unit.
- end panels required to cover the side of the refrigerator.
- valences over the cabinets or open to the ceiling. If cabinets are open to the ceiling, is there a finished trim allowed in the price to finish the top of the upper cabinets?
- corner cabinet for the lazy susan or shelves is angled or square to the corner.(see Figure 26-67).
- drawer locations, sizes and contents.
- height placement of all upper cabinets if the wall height is more than the normal.
- island or counter section to fit the cook top unit selected.
- glass or special cabinet door features.
- special shelves or nick knack areas.
- special spice shelves on the cabinet doors.
- correctly named the color, style, and design of the counter top.
- correct distance from the counter top to the under side of the upper cabinet.

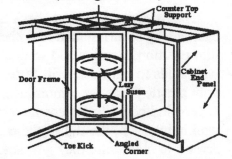

Figure 26-67. Lazy susan drawing.

- cutting of the counter top for the sinks, cook top with venting to the basement, and counter top mounted plugs included in the price.
- linen, pantry or broom closets supplied by the cabinet manufacturer. Include any pull out drawers or special shelf units in the price estimate.
- time to service damaged cabinets.
- toe kicks included in the price of the cabinets.

- pull-out cutting boards included in the estimate.
- narrow cabinet area for trays and breadboards.

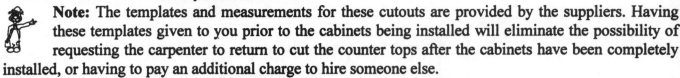 **Note:** The templates and measurements for these cutouts are provided by the suppliers. Having these templates given to you prior to the cabinets being installed will eliminate the possibility of requesting the carpenter to return to cut the counter tops after the cabinets have been completely installed, or having to pay an additional charge to hire someone else.

Cleaning Note: It is the responsibility of the carpenter to sweep, pick up, and remove all his debris from the house, and place that material on your trash pile for pick up.

Premanufactured cabinets are considered the most economical because they are built on an assembly line with standard modular widths and heights. They are usually built using a cheaper wood or pressboard, and laminated with a veneer wood or vacuum plastic finish. The cabinets are assembled by stapling and gluing the wood members together which will limit their durability and therefore their lifespan.

The different cabinet styles are exhibited in a showroom displaying the different kitchen configurations and designs with a wide variety of colors, finishes, faces and hardware styles. They are limited to fixed, rectangular, modular sizes which do not always allow the cabinets when installed to fill the space allowed for in the design. In order to fit, several different widths of spacers must be installed between each modular cabinet until they fill the required area. Visually they are attractive from the outside, but once the cabinet doors are opened there is less actual storage space than first

Figure 26-68. Illustration of a dead corner.

imagined. Rectangular and modular configurations create dead corner spaces, and due to the rectangular shapes do not fit on angled custom walls.(see Figure 26-68).

The cabinets, without the doors attached, are delivered by the carpenter to the job site in a large covered van. The cabinets are numbered in sequence of installation, and placed in the designated rooms to eliminate errors when installing. Inspect the cabinets and counter tops as they are delivered for any scratches, dints, discolorations, or questionable markings that will require their return to the manufacturer for servicing. If not caught on the delivery day or shortly after, it will be very hard if not impossible to maintain the overall construction schedule due to the delay imposed by trying to return any cabinets once they have been installed.

The carpenter will first check the floors for level. If the cabinets require leveling, he will attach the base cabinets by securing them together with wood screws to make sure they will not separate, and slide them into position. Using wood wedges he will adjust the height of the cabinets until level, and then locate the wall studs to which he will fasten the cabinets with long wood screws. The screw heads will be covered with plastic plugs for camouflage.

To find correct placements for the upper cabinets, the carpenter must first measure to find the 7 foot wall height, i.e., the top of the cabinet. Once again he will attach the upper cabinets together with wood screws to make sure they will not separate when positioned. Using adjustable supports the upper cabinets are raised into position, the wall studs located, and the cabinets fastened to the wall with long wood screws covered with plastic plugs. Wall heights may vary with the design of the house. If higher than the norm, make sure your cabinet designer and supplier are aware so that they can make the necessary adjustments. They might be required to provide higher upper cabinets so the visual proportion of the cabinets to the room will appear in balance.

The counter top supports and the finished counter tops are glued and precut to size in the shop except for one end. If adjustments are required the cabinet length is measured, and the counter top is cut on-site to fit the adjusted length. All perimeter joints, counter top splash backs, and cabinet attachment to the walls will be covered with a bead of caulking that matches the cabinet and counter top colors. Once the counter top is securely fastened to the base cabinet the carpenter installs the sliding drawers, door faces and hinges onto the correct cabinets. He then adjusts the hinges for a proper door fit, and installs the knobs.

One significant benefit of the modular unit is for the handyman willing to spend some time in the evenings and weekends to install the cabinets and save some money. Most modular cabinet manufacturers provide detailed instructions for installing their product, and these directions should be studied and carefully followed. As previously stated, floors and walls are seldom exactly level and plumb, therefore shims and wedges will be required so the cabinets are not broken or twisted during installation. Doors and drawers can not be expected to work properly if the cabinet framework is not installed correctly.

These cabinets however limited by their design, selection, and quality of materials have found their place in the new home market. If this home is your first starter home, and budget is number one on your list, consider the premanufactured modular units.

Custom built and pre-assembled cabinets have more selections in color, design and finishing. Because they do not have specific widths and heights, the cabinets can be manufactured to fit the required space with only end spacers required to finish the cabinet face. Inside every inch of space is utilized, dead corner spaces are non-existent, and because they are custom-made the cabinet shape will be able to fit onto any angled custom walls.

The cabinets are all manufactured at the cabinet shop, and delivered to the job site in numbered sequence for installation. As with the premanufactured units the carpenter must check the floors for level, and adjust the base cabinet level by fitting wood wedges underneath. Again the screw heads are camouflaged with plastic plugs. The carpenter will find the 7 foot wall height for the top of the cabinets, attach the upper cabinets together with screws, and position and secure the cabinets to the wall studs. (see Figure 26-69).

Most kitchen cabinet manufacturers order pre-fabricated counter tops from companies who specialize in these applications as it is the more economical route for the supplier as well as their clients. The counter top supports are assembled

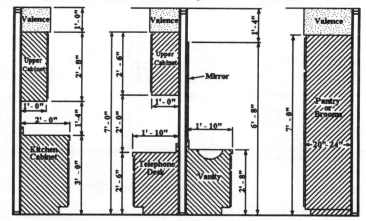

Figure 26-69. Cross section showing sectional view of typical upper/lower cabinet requirements.

in the shop without the finished counter tops installed, and then measured and cut to fit tightly to the wall and side cabinets on-site. Once satisfied of the fit, the arborite/formica sheets are glued to the counter support with contact glue. All rough edges are routered, sanded, and filed for a smooth finish. To complete the job the base cabinet, perimeter joints, counter top splash back, and the cabinet attachment to the walls are covered with a bead of caulking that matches the cabinet and counter top colors. The carpenter before leaving will do any required paint touch-ups, and install the drawers, door faces, and hinges on the correct cabinets. He then adjusts the hinges for a proper door fit, and installs the door knobs.

Custom built and pre-assembled cabinets are not limited by their design, however they are more expensive than the modular units. Because of their wide range of color and selection, and high quality of materials they are usually included in the specifications of custom home builders for the middle and upper end housing market.

Cabinets built on-site are considered the cadillac of all the cabinets. If lucky enough to find a carpenter that is able and willing to custom build your kitchen, bath and laundry cabinets on-site you will have quality cabinets that will last a lifetime. The price should be comparable with the custom built and pre-assembled cabinets.

Time factors should be considered when having a carpenter build the cabinets on-site. Compared to having them pre-built and semi-assembled in the shop, and installed on-site which only takes a few days, the carpenter might take a week or two. This time period should be scheduled into the construction time, and adjustments made for the house completion date.

Because you will be dealing with a carpenter and not a salesperson, he may not be able to do a detailed drawing, therefore, take sufficient time to thoroughly discuss and draw a sketch of the drawer, storage, and cabinet requirements. An alternate suggestion is to hire the designer to review your cabinet needs, and draw detailed floor plans and elevations of the cabinets for the carpenter. In order to do this you will have to provide your designer with the actual lengths and heights of the walls taking into consideration the locations of the windows, electrical switches and plugs. The size of the appliances should also be available along with any custom features such as appliances being set into interior walls in order to ensure they are accurately reflected in the design. (see Figure 26-66).

The carpenter will build the cabinets from the bottom up using plywood or paper templates, and will sketch locations of the appliances and cabinet parts on the floor and walls for correct locations and

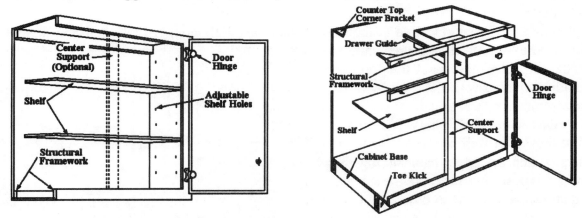

Figure 26-70. Illustrations of typical upper/lower cabinet assembly.

measurements. The base or toe kick area is constructed first and nailed securely to the floor and back wall after making sure that the floor is level. (see Figure 26-70).

Once the level has been found, the skeleton or structural framework of the base and upper cabinets is constructed including the installation of the end panels of the cabinets. Once the framework and panels have been secured to the walls, the perimeter face frames which hold the hinges for the doors and drawer guides in place are installed, and the shelves measured and cut for installation. Now that the basic structure and the rough opening locations for the drawer and cabinet doors have been determined, it will be neces-

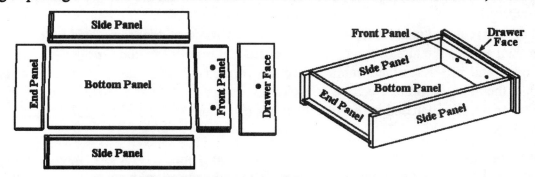

Figure 26-71. Illustrations of typical drawer parts and assembly.

sary to measure the size requirements of the actual drawer frame to be built and installed. The drawers are glued, checked for square, nailed together, and then set to the side until the glue has completely set.(see Figure 26-71).

The counter measurements are taken, and the counter top supports assembled without the finished counter tops. The counter supports must be measured, and cut to fit tightly to the wall and side cabinets on-site. Once satisfied that the counter support will fit, the formica sheets are cut to size and glued to the counter supports with contact glue. All the rough edges will be routered, sanded and filed for a smooth finish. When the glue is completely set the perimeter joints of the formica, counter top splash back, and cabinet attachment to walls will be covered with a bead of caulking that matches the cabinet and counter top colors.

The cabinet door and drawer faces can be made by the carpenter, however this will add about a week to the contract. Many carpenters order the faces from companies that specialize in these prefabricated items, and provide a wide variety of styles, sizes, and finishes, eg., painted or stained. These are delivered to the job site pre-cut to the exact dimensions with edges and hinge holes machined and ready for installation.

Assuming that the cabinets will be stained or painted by yourselves or the painting contractor, the cabinet maker will install the hinges onto the prefinished cabinet doors, and mount the door in the opening. He will install only one screw in each hinge, set the door in place, make the necessary adjustments to the door for a proper fit, and then set the remaining screws into place. The prefinished doors can then be removed in preparation for the painting of the cabinets. When the paint or stain has completely dried, the carpenter will return to re-install the cabinet doors, and mount the door knobs to the doors and drawers.

In personal experience I have had all three types of cabinets installed, but found that the custom built and pre-assembled units were preferred for 90 percent of the houses. The possession date for move in usually required scheduling the finishing of the house with cabinets, painting, carpets and finishing to a very tight time line. However, I remember the built-in units with much satisfaction and pride as all our cabinet needs were considered and installed as per the design. It was a joy to have every square inch used, and all the desired features at our fingertips. Because of this we have always said that when it comes time to build our last house the cabinets will be built on-site, and take however long it takes, without compromise.

♦ **FINISHING CARPENTER:**

The lumber supplier usually delivers all the finishing materials to the job site, and it is the finisher's job to cut and install the materials which give the house a more livable, completed look. Some of the most commonly used materials are fir, spruce, pine, and basswood. The wood selected for finishing should match as closely as possible the cabinet style and finish already selected. A painted cabinet allows finishing materials to be selected from paint grade lumber such as pine or basswood; stained cabinets limit selection to the finishing materials that will, when stained, match the color and grain of the cabinets. Oak is usually the most widely selected finishing material for the pleasing appearance of the grain, and its ability to hold many colors of stain.

The finisher will first install the plywood overlay over the rough tongue and groove subfloor in the areas that will need a smooth, finished surface. In areas where ceramic or quarry tile will be laid the finisher will install 3/8" good one side (G1S) overlay, and in areas where linoleum tile or sheets will be placed the finisher will install 1/4" good one side (G1S) overlay. These areas must have a clean, smooth, flat surface for proper adherence of the glue and cement. Bubbles or knots can cause future lifting and cracking of the tile, or unsightly blemishes which will be visible through the linoleum flooring.

The finisher will also install the door jambs to the rough openings of the doors. One door jamb consists of two side jambs and a head or toe jamb which when assembled will hold the door, the door casings, and the door stops in place. The jambs have already been routered to fit the doors hinges, and are usually precut for the wall thickness and door height. Some jambs may come pre-assembled with the door hinged and mounted to the casing, but without the door stop installed. If this is the case ask for a credit from the finisher because he does not have to do any of the work to assemble them. Before installing the door jambs the finisher must make sure the jamb is set square in the rough opening, and in order to do this wedges and spacers are used around the rough opening until the jamb is square and fits the finished door size. Nails are driven into the jamb through the spacers and wedges, and into the wall studs along the line where the door stop will be nailed. The door stops once in place will hide the nail heads. (see Figure 36-72).

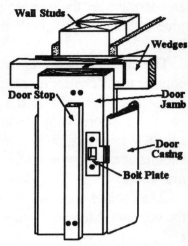

Figure 26-72. Door jamb assembly drawing.

The interior swing doors are usually delivered with a pre-drilled hole for the door knob, and side mounts pre-routered for the hinges. When ordering your door make sure that the door jamb and the door have been routered to accept the same number of hinges. Depending on preference, you can order the doors and jambs with two or three hinges. Review the blueprints, and make sure the swing and hinge location of each door is discussed with the lumber sales representative. This will eliminate having a door swing the wrong way, hit another door, or cover up the room wall switches.

Hollow core swing doors are most commonly used for interior doors because of their cost and light weight, and the options for stain or paint finish. A hollow core door is constructed with a perimeter frame carefully sized to fit your specific door opening, and an allowance for only a slight amount of on-site cutting and adjusting. The perimeter frame can also be ordered with a wood grain finish that will closely match the door face should you decide to stain and lacquer your doors. Over-cutting the perimeter frame will reduce the structural integrity of the door, and result in warping due to the inability of the inner core of softer and less expensive wood materials to withstand the pressures applied. The interior is composed of cardboard or soft wood baffles which support the wood veneers on the face of the door. Hollow core doors with designer panels are constructed by mixing the hollow core sections with molded plastic or wood fiber components to produce a very nice wood grain finish.

Note: When the door is installed onto the jamb and tested, the finisher must provide sufficient space between the door and the finished floor for proper air circulation, and adequate clearance for door swing. The proper space from the finished floor to the underside of the door for air circulation is no less than 1/2", and no more than 5/8".

When the finisher has installed the hinges to the jamb and set the door in place, he will then install the door casing making sure the entire door jamb is rigid. The next step is installation of the perimeter door stop to ensure the door will close comfortably, but not too snugly to the door stop. This step will allow a proper alignment of the door knob and strike plate to the hole drilled to accept the door latch. When closing the door the latch should catch the strike plate easily without requiring force to close the door. When the casing is installed to the jamb, the bottom of the casing is cut to accept the thickness of the finished floor under it. (see Figure 26-73 and Figure 26-75).

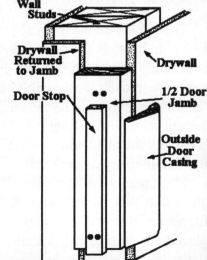

Figure 26-73. 1/2 Jamb drawing with casing cut to accept the finished floor.

Bi-fold doors commonly used for interior linen and closet doors consist of pairs of hollow core doors hinged together. The folding action is guided by an overhead track hidden and screwed into place beside a half jamb. Unlike a swing door which requires a full jamb because it will be seen when entering and exiting the room, a bi-fold door only requires a half jamb (see Figure 26-73), as it is usually only seen when the closet door unit is closed. The bi-fold doors are pre-assembled, and manufactured the same as a swing door to visually match the selected pattern and style.

The bi-fold door tracks hold and direct the center guides which have self-lubricating nylon bushings that ensure smooth, quiet operation. The weight of the doors is supported by pivot brackets and hinges between the doors -- not by the overhead track and guide.

A variety of door sizes can be selected. Two door units generally range from 2 to 3 feet wide while four door units are variable in total widths of 3 to 6 feet. When the four door width is wider than 6 feet, bi-pass or sliding door units with heavier hardware are suggested. (see Figure 26-74).

The cutting of the casing to accept the finished floor thickness determines the height of the baseboard to be attached to the walls around the perimeter of each room.

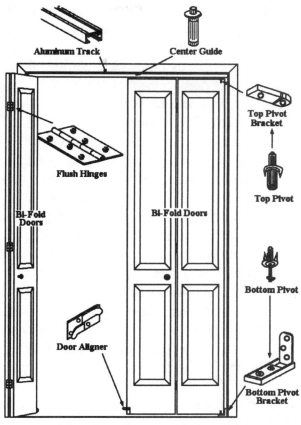

Figure 26-74. Illustration showing a typical 4 section bi-fold door with its components.

The baseboards are usually the same style as the door casings, but usually taller to give a more finished look between the walls and the floor. The finisher cuts the baseboards to fit, and leaves them in the house for the painter to stain or paint. Baseboards are attached to the walls before the carpets are installed; carpets are cut to be stretched and fitted under the baseboard. He will return to install the baseboard after the lino or tiles have been installed so that the baseboard can be nailed tight to the lino or tile, and eliminate any unsightly gaps. The thickness at the bottom of the baseboard must be wide enough to cover any gaps or floor undercutting caused by the installer. These gaps can become more obvious if the floor joists are set into place with the crowns arranged in different directions which can cause a wave in the floor's finish.

When the baseboard is nailed level with the door casing, there will be obvious spaces visible between the finished floor and the bottom of the baseboard. This can not always be corrected, but is usually not too unsightly. As previously mentioned, one way to minimize this problem is by making sure the finisher allows for the proper door casing adjustments, i.e., when one doorway has two different floor heights, the bottom of the casing must then be cut at different heights to accept the different floor levels. (see Figure 26-75).

 Note: Make sure that your finisher has been informed of the floor's finished thickness, i.e., lino, tile, or carpet. This way the casings or doors are not

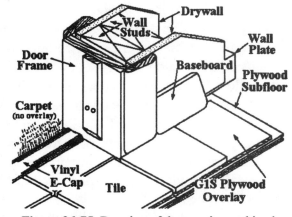

Figure 26-75. Drawing of door casing and jamb to fit different floor material heights.

over-cut possibly leaving a large gap between the finished flooring and the under side of the casing or door.

The casing selection for the windows should be the same style as on the doors so that they will match if located close together.

Any opening that requires covering with a frame and a finished casing are installed by the finisher. This includes service access hatches at the backs of tubs or spas, and attic access hatch boxes to be built, insulated, and weather stripped to keep the cold attic air from entering the house in the winter.

Because of their work finishers tend to be very temperamental and finicky. They work their own schedule, eg., not showing on the first day, or arriving late and working into the night, or some days not showing because another job was overlapping, or they just did not feel like coming in. They like to complain to anyone who will listen about the quality of the drywall finishing, etc., which tends to upset the owner's wife as she would like her dream home to be perfect. I always suggest that they show the builder and subtrade in question the problem area, and have them prove that they are unable to work around or repair it themselves. After that it usually gets completed without any more verbal complaints. It is my belief that everybody will do their best to make you the owner happy, and if not they can expect to receive some repair back charges, and not to be hired again.

It is in your best interest to get estimates from several finishers, and keep in touch with a few of them during the different phases of the house construction. Keep asking them how their schedule is going, and if it will coincide with your construction schedule. This way, if one of the finishers cannot show when you need them, you will have the option of hiring another.

The finisher, as his name states, must provide in his price estimate all the necessary labor, nails, staples, and glue to install all the materials that give the house a finished look ready for occupancy. His price will depend on the amount of work he will be required to do. The following list will assist you when requesting price estimates.

The finisher is to:
- sweep and prepare the subfloor for the installation of the overlay.
- nail and staple the overlay in all areas where linoleum, tile, marble, or any other flooring requires a smooth and finished surface for glues or cement to bond.
- build and install all attic access boxes with rigid and batt insulation, perimeter weather seal and casings.
- cut material and install with perimeter casing all plumbing service access hatches.
- square and install all interior door casings for all exterior doors, and caulk around the exterior of all doors for weather tightness.
- square and install all swing, folding, and by-pass door jambs, doors and casings.
- cut, drill, mount, and install all hinges, strike plates, door tracks, door knob hardware and locking devices for all interior and exterior doors.
- cut to fit and install all baseboards as required after paint or stain finish.
- sand all joints, corners, and nail holes filling them with sealer ready for stain or paint.
- install all window casings.
- build a custom, or install a pre-fabricated wood fireplace mantel.
- install all spring door stops to baseboards for swing doors.
- measure, cut, and install all linen, pantry and closet shelves, brackets and clothes rods.
- install all brackets, posts, balustrades, and stair handrails to the basement, main, and upper floors.
- measure and install all bathroom towel bars, paper holders, soap dishes, or grab bars as required.
- install all built-in units such as medicine cabinets, ironing boards, etc.
- install all garage door closure devices required by building code.
- measure, cut and install additional weather stripping as required for exterior house and garage doors.
- install additional storm or screen doors to house.

- install any custom shelves, window features, or hardware as required.
- cut and fit material for wood feature walls.

Cleaning Note: It is the responsibility of the finisher to sweep, pick up and remove all debris from the house, and place that material on the trash pile for pick up.

◆ PAINTING CONTRACTOR:

The painting contractor will be required at the job site during different times of the construction in order to prepare and finish his work. It is his job to protect from moisture, and enhance the appearance of the interior and exterior surfaces of the house. Most painters suggest a latex paint for the walls because it hides many faults and blemishes that would normally be visible in the drywall, and it provides a non-glare, low maintenance finish to the walls. The doors, trims, baseboards, and casings require a higher maintenance finish than the walls, so an oil base paint is often used. A simple soap and water mixture will remove most stains and finger prints from oil base painted wood surfaces. Choose from hundreds of colors and tones of paints, stains and other finishes manufactured for interior and exterior applications. When the wood surface is properly cleaned, sealed, and painted the better quality paints or stains will last a good four to seven years.

The quality of the paint selected by the builder and the painting contractor will determine how easily it can be cleaned and maintained by the family. Try to choose good quality paints and stains, especially since the cost of the materials is usually only 10 to 15 percent of the total cost of painting. The painting contractor's price estimate is usually determined by the quality of the paint or stain selected, and the amount of time required for painting or staining the walls, trim, and woodwork. Most painting contractors charge a cost per square foot of wall area to be painted once they know your requirements.

The biggest mistake made when collecting estimates from the painting contractor is to forget to include the proviso "to clean all surfaces to be painted by removing any dust or chemicals that will interfere with the adherence of the paint or stain". If he is a good painter and cares about his work, he will do the cleaning automatically. However, just in case it is overlooked, have the cleaning of the surfaces put down in writing as a part of the estimate.

As discussed on page 138 of this chapter, the primer coat will be applied to the interior walls quickly, usually taking a few hours, by using a sprayer or roller. This occurs after the drywall taper has completed sanding, and before the ceiling texture has been applied. The subcontractors are used to this scheduling, so you should not have any problems keeping the texturer out while the painter is priming the walls. In fact, the person in charge of drywalling or texturing will most likely remind you to schedule the painter.

Some time has passed since the painter was last at the job site, and during that time many suppliers and subcontractors have been through the house dinting, marking, and scratching the walls with tools and other items. Hopefully, under your supervision they have kept the damage to a minimum. Now before the painter can apply the first finished coat, these defects must be repaired using a filler. Once dry he can apply the first coat of paint to the walls and woodwork. He uses a brush to apply a thick coat of paint to all the door casings, trim, and wall and ceiling corners. Rollers are not able to reach those hard-to-get areas, so when the painter uses a roller to apply the paint to the walls these corners are already covered.

The baseboard that has been cut and set aside by the finisher is painted and left to dry. The finisher will return after the finished flooring has been installed to complete

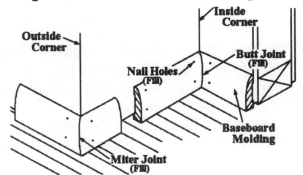

Figure 26-76. Illustration showing holes and corners to be caulked prior to painting.

the baseboards. The section of the baseboard to be installed over carpet must have a bead of caulking applied to where it touches the walls to produce a smooth and finished look to the wall. Caulking to the baseboards will be applied later by the painter once the finished flooring has been completed throughout the house. Once all the baseboards and flooring have been installed the painter will apply the caulking and filler to all the remaining corner joints, and cover them with a coat of paint. (see Figure 26-76).

During this time the electrician has been back to install all the lighting and switch plates. Because he is working with items that have to be installed on painted surfaces, he will most likely scratch or mark some of these surfaces. When the painter returns to the job site to paint the baseboards, they will patch any scratches or marks on the wall, and apply touch-up paint to those areas. This is the last time the painter will repair any damage done by other subcontractors so make sure they touch-up every wall, trim, or door blemish before leaving.

If you have requested the application of two coats of paint to the house, they will have to apply the second coat after a full 24 hours of drying time with the house temperature at a constant 20 degrees centigrade or 68 degrees Fahrenheit. Some of my clients ask if a second coat of paint is really necessary because of the initial application cost. Usually during the first winter and spring the structural teleposts will have to be adjusted because of the house foundation settling. This settling may cause the drywall nails and joint tape to crack. During this time the structural wood members in the walls, floors, and ceilings will also dry out and shrink due to the winter heating requirement which causes similar drywall cracks and popping nails. Sometimes the corners for the door and window casings separate during settling, and these gaps require filling and sanding before the painter applies his second and final coat of paint. Therefore, I suggest a delay on the second paint coat for another year if the owners do not mind the job of moving the furniture, and the painters removing all the electrical wall plates and pictures.

Exterior wood surfaces must be protected with a good quality oil base stain. Special care must be taken to clean the surfaces, and remove any dirt or oil so that the sealer can penetrate into the pours of the wood, and eliminate any moisture penetration. The sealer is usually an oil base paint or stain for maximum protection. The sealer color is usually the same as the finished coat, and once dry the finished coat of stain or paint is applied. If there are any exterior touch-ups required, the painting contractor will do this at the same time he returns to complete the interior touch-ups.

During the life of the house the owner will be required to do some repainting of rooms, or some touch-up painting. Have the painting contractor leave at least a gallon of each type and color of paint used in the house, and mark the containers with the re-order numbers for future reference, and the room for that particular color.

Cleaning Note: It is the responsibility of the painting contractor to pick up and remove all paint cans, used brushes, and solvents from the house, and take them to the dump, or city chemical and waste reclamation center.

♦ FINISHED FLOORING - CARPETS, LINO, TILE, etc.:

The term "finished flooring" applies to the material used as the final wearing surface of the floor. A wide selection of materials are prefabricated for this purpose.

It is strictly a matter of personal taste in the selection of floor finishes. It is important to note that when selecting the flooring for each area that it is not only price, appearance, and wear, but the type, density, preparation and installation of the underlay material that will determine how the floor will look, and how long it will last. The flooring installers must sweep and prepare the subfloor or plywood overlay by nailing, filling, sanding, and sweeping the floor area before they install the finished flooring. The installers will supply all the materials necessary to install the finished floorings, however it is very important that the plywood overlays are firmly secured to the floor by the finisher. It is also important to have included in the estimate, that the carpet installers will return to re-stretch the carpet after one (1) year without charge.

Today, a tremendous selection of floor coverings is available. Choose from elegant plush carpets, multi-colored, sculptured designer carpets, artwork area rugs, decorative resilient sheet flooring, or natural materials such as prefinished woods, tile, stone, slate, or brick. Keep in mind that patterns, tweeds, and small floral patterns will conceal dirt and footprints better than a one-tone light or dark colored carpets, but if you wish the latter, relegate them to areas of least traffic such as bedrooms.

Before making the final decision on the floor covering to purchase, consider the maintenance that will be required given the traffic patterns,and use of that area of the house. Make sure that the floor finish chosen has a high resistance to water, will not soil easily or show dirt, and will be easy to maintain.

In addition, consider the following when making floor covering choices:
- cushioned resilient flooring or carpeting is a foot saver.
- avoid hard surfaced flooring in areas that require a lot of standing such as the kitchen and laundry areas.
- avoid area rugs in areas of the house used most by children or the elderly, because they are easy to trip over.
- carpets and cushioned flooring increase the insulation factor, and carpets help to cushion the noise between floors.
- many carpets and resilient coverings are manufactured specifically to resist the moisture and molds that are present in the more humid areas of North America.

To unite rooms or to expand small areas use the same floor covering throughout. A floor covering in a solid color creates an illusion of space, while a pattern gives a feeling of intimacy. Geometric designs give architectural elegance to rectangular rooms, while random designs disguise irregular room shapes.

When first going to the supplier's show room, take along the sample paint colors as well as fabric samples of furniture, draperies, and venetians. A color slightly darker than the walls is a safe choice.

Carpeting:
Wall-to-wall carpeting provides comfort under foot, and works well with traditional and modern furnishings. Wall-to-wall carpeting is found in every color imaginable at almost every price. Sophisticated dying techniques make colors fast.

The keys to quality are fiber content, the way the carpet is manufactured, and the density of the pile. The density and height of the pile or surface yarns determine wear and quality. A good quality carpet is a balanced combination of pile height, pile density (closeness of yarns), and ply (number of strands) of the individual yarns. High density means less abrasion on the sides of the yarns, a richer, fuller look, and less footprinting. For very heavy-traffic areas choose carpeting made of tight twist, high-ply yarns. Test the density of the carpet by bending the fiber side of the carpet toward you.

If you have never experienced it before, you may be horrified during the first few weeks of vacuuming a new carpet to see balls of fuzz and sprouts appearing. Do not panic as this is perfectly normal. The vacuuming will pull up excess carpet fuzz, and clip never pull sprouts. Both phenomena should stop after a few weeks.

Underpad cushioning:
Regardless of the quality or luxuriousness of the carpeting or rug purchased, make sure you also select the proper underpad. Proper cushioning is essential toensure a long life for the carpeting. In addition it helps protect your feet, insulates, and helps muffle sound transmission.

When selecting the undercushion, keep the following points in mind. Thick, lush cushioning is best for low-traffic areas in the home. The thicker the cushion the more the wear to the carpets. Waffled construction in buoyant foam rubber gives a rich, luxurious feeling; sponge rubber is resilient and soft under foot, but will hold moisture; urethane cushioning is resistant to heat and dampness. Firm under-padding is best for high-traffic areas. The synthetic material will resist mildew and make carpets last longer. This type of cushioning is usually installed under carpets in most commercial sales areas such as the showroom of the supplier. The carpet supplier should be able to recommend the best underlay for your carpet selection.

Sheet flooring:

This type of linoleum flooring is installed by cementing it to a smooth plywood overlay. The 6', 9' and 12' widths give a smooth, seamless, wall-to-wall effect. Because of the seamless appearance and resilience, sheet goods provide an ideal answer for areas exposed to surface moisture such as laundries, washrooms, mudroom entrances, and kitchens. The cushion resilient goods are especially effective in areas where you stand for relatively long periods of time such as the kitchen and laundry areas.

Care of resilient flooring is important. Use only the prescribed cleaning agents, or a vinegar and water mixture. Some resiliences can be damaged beyond repair with harsh, strong cleaners. Dust daily and wipe up spots immediately. For the ultimate in convenience, try the no-wax resilient which needs only regular wet-mopping to keep it maintained.

Hard-surface flooring:

As discussed earlier, hard surface flooring can be wood, ceramic, brick, marble, or slate. When used in very high-traffic areas, such as foyers, they will require special care, but the beauty will more than compensate.

- *Sliced brick* can be country or contemporary, and lends itself to indoor/outdoor areas such as spa or pool decks.
- *Ceramic tile* comes in many glazes, patterns, colors, and shapes, and is easily mix-matched. The grouting is also available in a variety of colors.
- *Marble* is very elegant, and comes in various sizes and colors. It stains very easily, is very expensive, and is hard to maintain so is used sparingly.
- *Quarry tile* is fired from clay, come in a wide selection of surface textures, and is available in their natural colors, i.e., brown, rose, and slate blue.
- *Slate* arrives pre-cut in many colors and thicknesses. If you have traveled the mountain highways you have most likely passed by many color variations of slate including black, purple, blue-grey, charcoal, rust, green and several brown tones.
- *Terrazo* is marble or stone chips set in mortar, then polished to a smooth, shiny finish in a multi-colored, multi-shaded blending. This style of marble flooring is usually found in entrances of commercial high rises with a very high traffic flow.
- *Wood flooring*, especially hardwood, has the strength and durability to withstand wear, and provides a highly attractive appearance. Oak is the most widely selected, however maple, birch, beech and other hardwoods are also desirable. There are three general types of residential wood flooring: strip, plank and block. Wood flooring must be the last in the the sequence of the interior finishing steps. Manufacturers recommend that wood flooring be delivered four to seven days before installation and piled loosely throughout the house. This permits the wood to equalize its moisture content to that of house.

♦ GARAGE DOORS:

Overhead garage doors come in all heights with many widths, styles, and designs. Sectional overhead garage doors are the most common door sold in the residential marketplace. They are called sectional because they arrive pre-built in sections, and then each section is hinged to the inside of the door to allow for ease in the vertical roll. The sectional door can be constructed of almost any type of material, but the two most common are wood and metal.

The metal door is light, and is considered almost virtually maintenance free. Although slightly more expensive, it has become the most common as it is lighter in weight, and better insulated than a wood door. The metal door is usually constructed with two molded, metal veneer faces, and filled with a polyurethane insulation giving it a very high insulation value. Metal doors can be ordered with a wood grain finish.

The cedar sectional door, although the most expensive of the wood overhead doors, is used mainly by those people who would like a visible wood finish. The beauty of the cedar grain is realized and protected with a good oil or stain finish which prevents the acceleration of weathering for many years. For this reason the construction of the wood door is limited to only a few types of wood as some woods will warp with certain weather conditions. Wood doors can now be purchased with a highly weather-resistant, plastic or paint finish which will make them cheaper than cedar in price, and will provide the home owner with many years of maintenance free operation. A wood door is much heavier than the metal door, and is considered insulated as wood acts as its own insulator.

Both metal and wood sectional doors have the advantage that they do not swing outward, and therefore allow you to park the car closer to the garage door itself. Sectional doors are best suited for an electronic door opener since the mechanism will not have to work hard as each section is usually designed with an easy roller at each side. Because of the weight difference a less powerful door opener is needed for metal doors which may make them comparable in price to wood doors.

If you own a van you should consider purchasing a taller door unit. The most common door height for a van is 8 feet, although a higher door yet should be used if an air conditioning unit is mounted on the roof.

The American and Canadian Governments have ranked the top three causes of heat loss via garage doors. The most obvious is leaving the door open while the garage is not in use. Air infiltration between the sections is the next major source of heat loss. This is compounded if the door does not have a proper seal and perimeter weatherstripping. The greatest misconception is that a properly insulated door is the more important factor for an energy efficient door. Studies show that the door does not have to be thick or well insulated, but rather must have proper sealing or weatherstripping. Ensure that the door purchased is adequately weatherstripped on the bottom and the sides. Vinyl or rubber are most commonly used as they are the most flexible, and tend to withstand the changing weather conditions. The top of the door is usually sealed with a wood or metal stop. All insulation and weatherstripping should be added to the door unit by a professional installer at the shop prior to installation. This will guarantee your warranty, and eliminate any problems that might arise from the do-it-yourselfer.

Convenience... Security... Energy Savings. These are the major reasons home owners decide to buy an automatic garage door opener. Good quality openers certainly justify the investment. Coming home on a snowy, cold, and dark night will never again be an adventure into the unknown. Most people think bargain hunting for a garage door opener is a simple matter of saving on the sales cost, the installation cost, and installing it in an afternoon. Consider that you will probably have to rebalance the door before installing the opener, and this could cause more loss of time and money than it is worth. Buying items of this type on sale will usually only provide the basics, and things like an extended warranty, service, adjustment, weatherstripping, lubrication and extra transmitters will often be extras.

Garage doors and children are always a concern. Children should be taught and reminded that garage door openers are not toys, and playing with them could lead to serious injury. Ask for a safety reverse mechanism on the opener, and check it every six months to make sure no one will be trapped under the door.

♦ HOUSE CLEANING AND GARBAGE REMOVAL:

This job however dirty must be done on a regular basis during the construction of the house. The dirt, bent and rusty nails, soiled papers or rags, partially eaten lunches, half-full pop cans, lumber, and other scraps will accumulate in the house, or be spilled and ground into the subfloor. This material can potentially work its way up through the underlay, and stain or discolor the finished carpet or floor caulking. There are certain times during the construction period when it is advisable to thoroughly clean out the garbage and dirt that accumulates after each subcontractor has completed his work. The questions are, how often should this be done, and should you do the cleaning yourself, or hire someone on a contract basis to sweep and clean the interior of the house, and take the debris to the dump?

The house should be swept out weekly as regular maintenance. If this can be done personally in a few hours on the weekend, it will keep the interior relatively clean as well as the interior sweeping and trash removal costs down leaving the exterior trash pile removal as the only expense.

The house should have at least four thorough sweepings: after the framers, after the drywall tapers, after the texturers, and after the finisher. The fourth cleaning, after the finisher, but prior to the application of the finishing cost of paint is the time to rent a commercial vacuum unit which is capable of picking up small dust particles from between the subfloor and overlay joints without the hose and filters constantly obstructing. Its suction motor should be powerful enough to pick-up small nails, wires, staples, papers, and wood chips from the floor and heating ducts. Be warned that the residential shop vacuum units are not strong enough to do this job. The filters will constantly clog up and burn out the motor, and the suction is insufficient to pick-up small nails and staples. Believe me, this is experience talking!

There are companies in business exclusively to clean the interior and exterior of houses by sweeping and picking up the garbage, and taking it to the dump. They will contract to come to the job site 3 or 4 times during construction to clean and remove the trash from the house for a fixed price. They will also contract to do this work only on request, and the rate will be determined by the amount of trash they remove each time, eg., a full load, 3/4 load, 1/2 load, etc., with an added charge for sweeping.

Subcontractors are constantly running in and out of the house several times in one day, and the floors accumulate dirt very quickly. During rainy periods, mud tracked in will become layered which can only be removed with a scraper. If this is the case, it would become too costly to have a cleaning crew remove the accumulation of dirt and trash. The best way of keeping this expense to a minimum is to have each contractor include the cleaning of his materials and mess in their estimate. If after each day they pick up their trash and unwanted material, and throw it outside on a designated spot or a trash container, the dirt and trash collected inside the house will be kept to a minimum.

 Note: Trash piles should be removed after the framers are completed, after the taper has finished, after the finisher is completed, and after all the appliances have been installed and the cartons thrown out.

To avoid the garbage collecting on the property, a trash container can be rented on a monthly basis. These companies will provide a container that will be automatically picked up and emptied every month, or as required. These companies charge by the container size, and the number of times required to empty the container during the construction period. One negative to this method of collection is that some of the materials such as sheet metal, some plastics, and other trash that do not decompose quickly will not be accepted by a dump site. If these materials are thrown in, they will have to be removed before the container can be emptied. The subcontractors should be warned of this stipulation if you choose this method of garbage removal. The other negative side of this service is the ease of access for cheating by other trades and subcontractors working on someone else's house. Its very easy for a trades person with a half ton pick up truck to dump his garbage into your container after dark rather than take it out to the dump. You may end up paying for someone else's trash pick up as well as your own. This ease of access can be minimized by placing the container in a highly visible traffic area, or have the electrician wire a temporary flood light with a motion detector to the side of the house and aim it at the container.

NOTES

_____

_____

_____

_____

_____

_____

_____

_____

_____

_____

_____

_____

_____

_____

_____

_____

_____

_____

_____

_____

_____

_____

_____

_____

_____

NOTES

CHAPTER 27

• HOUSE MAINTENANCE:

Follow-up after the move-in date is an extremely important step of the building process. Proactive house maintenance requires only a few hours every 3rd or 4th month, and includes a walk-about the interior and exterior of the house to check for changes in the walls, trims, basement walls, structural posts, and perimeter settling of earth against the foundation. All these items can be easily noticed, and should be serviced as quickly as possible.

If experienced and good subcontractors or suppliers have been hired and provided quality materials, and all have paid special attention to the construction details, building code requirements, and information in this book, the house will require minimal maintenance compared to a house that is not well constructed and contains poor materials. Having a well built house of quality materials will reduce the cost and time required for maintenance, but does not eliminate the need to inspect and service during the first and succeeding years.

It is common during the first year in new homes for interior walls to develop cracks, have nails or screws pop, and have doors shift so they do not close properly. Also, floors may creak, and walls make loud snapping sounds during the day and night. As discussed earlier on drywalling in chapter 26, these problems are usually first noticed during and after the winter heating of the house. Wood members will shrink or shift because of the moisture evaporating with the heat; foundation bearing walls and structural posts sometimes settle at different rates. When these sounds or visual cracks first occur it is a warning that it is time for maintenance and service work. The drywall cracks, and nail or screw pops are under warranty, and should be serviced one year from the work being completed by the drywall contractor. Quarterly adjustments of the structural teleposts will reduce the damage to the drywall caused by the shrinking wood members and structural settling.

Inspection of the backfill material around the perimeter of the house will show settling due to the weight of rain and snow moisture in the earth which forces the backfill to compact downward. If not corrected within the first year by tamping and filling the voids, the water could pond and possibly freeze against the foundation wall. This puts unwanted pressure against the foundation causing the concrete to crack, or the perimeter backfill to settle too quickly. Providing a grade with a good slope to drain away from the house will eliminate much unnecessary service work in the future.

If you as the builder/owner develop and implement a well planned program of maintenance checks with occasional service work during the life of the house, the cost of upkeep will be reduced, the life of the structure will increase, and most importantly the value of your property will increase.

NOTES

CHAPTER 28

● PRESERVED WOOD FOUNDATIONS:

This construction method, although unfamiliar to many custom house builders, has become very popular with the do-it-yourself builder. Easy to construct in all weather conditions using only the framer, it provides a livable basement environment partly or entirely below grade level. The preserved wood foundation is a definite construction time saver compared to other foundation types because it goes together quickly in all types of weather, and does not require any special framing crews to erect the structure. Wet, muddy or frozen ground has little effect on the installation process which means that builders considering wood foundations will enjoy a longer building season without scheduling problems or weather shut-downs. It is suggested that all contractors considering an installation of the wood foundation system read the manuals provided by the local Wood Foundation Council, and the Government Building Standards Branch. This will eliminate errors in material selection, and construction problems that might occur during the construction period. There is really nothing complicated or unusual about the preserved wood foundation system as it is just an innovative engineering adaptation of a proven frame construction technique. Preserved-treated wood has a long history of outstanding performance as a structural support material in residential and commercial structures. A preserved wood foundation is a strong, durable, and contemporary building material that has a number of unique advantages over other foundation systems for the contractors and framers that build them, and the home owners who live in them. The energy-saving qualities of a wood foundation is one of the major reasons for their dramatic rise to popularity in North America. A wood foundation is

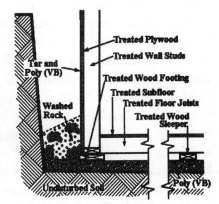

Figure 28-1. Drawing showing a wood foundation with wood floor.

efficient in terms of energy consumption because the entire foundation wall below and above the finished grade is totally insulated during the construction stage. The preserved wood foundation system is its own frost wall, whereas concrete foundations require the addition of a framed and insulated frost wall which adds cost to the construction, and reduces the actual living space in the basement. The exterior treated plywood and wall studs have natural thermal qualities (much higher than concrete) that are increased with the addition of R 20 to R 32 batt insulation between the studs making the entire perimeter of the foundation an extremely efficient barrier to heat loss. The wood foundation will require less energy to heat or cool the living environment, so fuel costs will be substantially reduced especially during the cold winter months.

Wood foundations have been used for about 25 years for single family, and low rise multi-family dwellings. They can be built with a conventional concrete footing and floor, or entirely with treated wood. In the latter, the wood footing and wood floor on sleepers usually sits on a granular bed of washed rock. All materials used for the construction of wood foundations must be treated under pressure with chemical preservatives to prevent decay. The high level of concentrated wood preservative makes the wood highly resistant to decay and insects that might destroy the structural integrity of the wood. (see Figures 28-1 and 28-2).

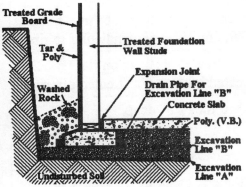

Fugure 28-2. Illustration showing a wood foundation with concrete floor and footing.

The wall structure of the wood foundation must be engineered to carry the vertical loads of the house and roof as well as the horizontal pressures of the backfill material. The engineer, knowing the requested backfill height and the number of stories the structure will support, must determine the size of the supporting studs, thickness of the exterior plywood sheathing, and distance of the stud spacing for the perimeter foundation walls. (see Figure 28-3).

The preserved wood foundations are constructed following the same proceedure for framing a standard house. The treated wall studs are nailed to a treated top and bottom plate which rests on a concrete or treated wood footing. The exterior wall which will hold back the foundation backfill is made of treated plywood sheathing with a spray tar coating, and a polyethylene cover that acts as a dampproofing material for the foundation sitting below the grade level. The space between the treated studs can be filled with batt insulation, and the interior walls finished to provide a full and warm living environment.

For future basement development on a treated wood floor system, the wood floor will give on weight bearing the same as the main floor joist system, therefore the special carpet underlay for concrete floors will not be required.

The manufacturer of treated wood materials provides a 30 year guarantee, although the wood material after this process will often have a life expectancy greater than the guarantee period. Pressure treated wood is environmentally clean and safe. The chemicals used in preserved lumber and plywoods are permanently fixed in the wood fibers, and cannot be leached out even under conditions of extreme moisture. Air quality tests have also been conducted and indicate that the air quality in areas where preserved wood has been used is equal to air enclosed by other foundation systems.

There are many requirements for a wood foundation that have been set by the Local building codes. However, there are options for using concrete depending on the local soil and ground water conditions. The following list will provide suggestions for the many options available when planning a wood foundation.

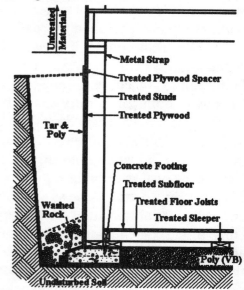

Figure 28-3. Illustration showing a concrete footing with treated foundation components.

	Concrete with Treated Foundation	**Treated Wood Foundation**
Footing Base	Undisturbed soil	Washed rock - 6" deep
Material Extension Beyond Footing	Undisturbed soil	Washed rock extends 24" beyond wood footing
Footing Materials	Typical - 16" x 8" concrete	2" x 8" treated lumber
Footing/Wall Sealant	Mastic gasket or Elastomeric caulking	Not required
Structural Pads	Typical - 30" x 30" x 8"	2 layers of typical 5 -2"x 6" treated studs on flat. 2nd layer perpendicular to first
Basement Floor Material	3 1/2" concrete	2" x 4" treated sleepers on flat with 2" x 4" treated joists + 1/2" treated plywood
Basement Floor Base	Washed rock - 8" deep	Washed rock - 6" deep
Foundation Wall	2" x 6" or 8" treated studs with 1/2" or 5/8" treated plywood	2" x 6" or 8" treated studs with 1/2" or 5/8" treated plywood
Foundation Wall Sealent	Elastomeric caulking	Elastomeric caulking

TABLE CONTINUED ON PAGE 165.

TABLE CONTINUED FROM PAGE 168.

	Concrete with Treated Foundation	**Treated Wood Foundation**
Additional Wall/Joist Supports	Hangers at all foundation openings, and metal strapping at all floor joists	Hangers at all foundation openings, and metal strapping at all floor joists
Waterproof Membrane Below Grade	Two coats spray tar + 6 mil poly - 8" above grade	Two coats spray tar + 6 mil poly - 8" above grade
Sump Pit	Treated wood or plastic	Treated wood or plastic
Wall-to-Footing Fastener Requirements	3 1/2" concrete nail every 48" o.c.	3 1/2" hot dip galvanized nails
Above Grade Fastener Requirements	Metal strapping at floor joists - 3 1/2" hot dip galvanized nails and staples	Metal strapping at floor joists - 3 1/2" hot dip galvanized nails and staples
Additional Requirements Above Foundation Wall	Typical untreated framing materials	Typical untreated framing materials

NOTES

_____
_____
_____
_____
_____
_____
_____
_____
_____
_____
_____
_____
_____
_____

NOTES

CHAPTER 29

• LANDSCAPING:

The landscaping of your property should take as much thought as the planning for the design and layout of the house. The land value usually represents between 15 to 30 percent of the total investment so it will be necessary to balance the values of the property, land, and landscaping into their appropriate percentages of expenditure. To do this effectively be prepared to spend about 20 to 35 percent of your property value, or 10 percent of the house value on the landscaping. This percentage should include the rough and finished grading, the walks, driveways, trees, shrubs, sod, and fence package as well as the sculpting and burming required for the lot to conform to the family's outdoor living requirements.

The house design and lot landscaping should be planned in conjunction with each other. The final value, enjoyment, and cost of maintaining your property will depend upon the placement, design, and quality of work and materials in the visual features of your landscaping such as driveways, trees, lawns, and fences. If the quality of materials and their construction is poor, their maintenance will become costly and time consuming. Structural and visual deterioration will eventually result in real property devaluation, and you will no longer have that quality streetscape once dreamed of.

Landscape planning involves:

1) - Preparing a detailed design for progressive development of the landscaping over a period of months or years to accommodate your budget.

2) - Hiring a bobcat or small grader to shape the land for proper drainage in relation to the house, and take advantage of the contours and different yard landscapes possible from each window location.

3) - Taking into consideration the installation and location of necessary features, such as retaining walls, driveway, sidewalks, lawn areas, terraced decks or land, and fences or hedges required for the different outdoor living activities.

4) - Selecting the proper seasonal trees, shrubs, and plants to complement the landscape design and the house's streetscape.

5) - Estimating the weekly and seasonal required maintenance for lawns, garden and permanent features to keep the property in good order and an enjoyable state over the years.

What factors to consider?

The following factors may affect landscape planning and should be considered:

a) - *Existing Grades.* Wherever possible take advantage of the natural features of the property such as rocks, tree stumps, existing trees, uneven grades, etc. As discussed earlier it is much easier and cheaper to build and develop a flat or level property, however a sloping lot or variable grade will offer the owner or landscape contractor a greater challenge for different and more interesting visual effects. Land that is above the street grade is preferable and easier to work with than low-lying ground where water can pool creating major drainage problems.

b) - *Existing Features.* Assess the property or adjoining properties for focal features such as a well treed backyard, a view, or an open field or utility corridor.

c) - *Climate.* Many northern Provinces and States are limited to about 4 months of warm temperatures or outdoor enjoyment. To make the best of this short season, plan decks or patios in sunny areas of the property, and sheltered from the wind. This way their use can be extended to include the early spring and late fall. These sunny areas are well suited for early growth of spring flowers and extended growth during the fall. Shrubs and plants located in these sunny areas will bloom two or three weeks earlier than those located on a northern exposure.

d) - *House orientation.* As discussed in chapters 3 and 5, it is important to locate living areas in a direction that will provide the best view of the landscape, and optimal orientation to the sun. The view and door access from each room to an exterior deck or garden will determine their location within the living environment. The smaller the lot and house, the greater the need for careful house and landscape planning as each room will want to share in a pleasing view of the property.

e) - *Family considerations.* If the backyard's landscape design is for the purpose of viewing a decorative display of trees, shrubs and visual plantings, it will have to be more carefully planned than a back-yard primarily to be used as a playground for young children. Each individual family member's hobbies, lifestyles and personalities will have to be considered when developing landscape design.

Grade Preparation:

During house construction, the rough grade with an adequate slope to drain from the house is completed by the excavator. Consult with the city inspector, house designer, landscaper or surveyor when the excavator backfills around the perimeter of the house as they will be able to determine if there is sufficient fill material against the foundation to allow for most settling situations. The first year of settling will result

in different ground compaction conditions around the lot and house. It is essential to direct all surface water away from the house. For this reason the finished grade at the base of the house must be high enough to allow for a slope down and away from the house. For this purpose, the excavator leaves sufficient backfill material on-site to fill any areas that may settle around the house perimeter. During preparation of the finished grade, sufficient black dirt will be brought in by the landscaper to fill these grade variations and satisfy the

Figure 29-1. Illustration showing extreme grades requiring retaining walls.

landscape design. The black dirt is spread in various thicknesses to comply with the finished grade and ensure there will be no ponding or erosion problems created. (see Figure 2-2).

If the lot's contour slopes at right angles to the street, the lot may require some terracing. Terracing must be carefully done in order to allow the surface water to drain away from the house and lot, and not cause potential flooding in the neighbor's lot. If the lot contours run parallel to the road, the yard slope may be too steep for a lawnmower. This contour type might require terracing via different levels with retaining walls to hold back the earth. The finished height of city road will determine the lot grade steepness and options for terracing. In this style of subdivision, it is common for the grade of the house opposite to be at the same height or only slightly higher than the city road height. In these situations it will be necessary to select gutter downspout and driveway locations to prevent water runoff from causing flooding or drainage problems for those neighbors that might be effected. (see Figure 29-1).

Whatever the site grading condition may be, the most important consideration in preparing the site is establishing finished grade levels at all corners of the lot in relation to the adjoining properties. The plot plan from the land surveyor will provide that information.

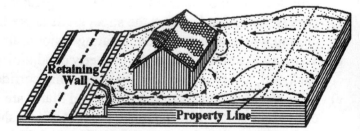

Figure 29-2. Illustration showing proper lot drainage.

Existing Topsoil and Trees:

If the subdivision developer has not stripped the topsoil from the lot consider yourself lucky. If the quality of the black dirt on the lot is good, have it scraped off and saved before starting construction. The excavation contractor will scrape the top 5 or 6 inches of topsoil off the area to be excavated, and pile it on a corner of the property. Stripping and replacing the topsoil will cost less than having that same amount of black dirt trucked in at a later date. If unsure of its quality, send a sample to the local Government Department of Agriculture for analysis, and request recommendations for the type of fertilizer required to enrich it.

Many developers when grading and stripping the subdivision topsoil have the trees cut down and removed so very few lots in the newer subdivisions have existing trees remaining. Some city planning departments are now requiring developers to spare the young and healthy trees around the lot's perimeters that will not interfere with the house construction. If the lot has protected trees, thin them out by selecting the healthy ones, and removing any dead or scrub bushes that might slow or interfere with the tree's growth. Young healthy trees should be saved, but consideration should also be given to their location and size when full grown. Any trees that will be too close to the house foundation should be completely removed to eliminate any future root damage to the house structure and drainage systems.

If the lot is without trees, it is worth the investment to plant several full grown trees to provide shade and immediate visual landscaping. These trees are relatively inexpensive; the tree spade that digs the hole and plants the tree is the expensive part of the process. Instant trees will be an immediate asset to the value of your house and property.

When selecting trees for planting it is best to view and purchase them from a local and reputable tree farm, i.e., one which has a wide selection of full grown trees and shrubs. Avoid those trees that tend to seed in quantity such as chestnut and silver maple, or those trees that sprout and support secondary ground growth as these seedlings will have to be constantly pulled and destroyed. Avoid planting types of trees such as poplars, willows, or elms too close to the house foundation as they have a heavy root system which can cause foundation damage and slow plant growth. These types of root systems also find their way into nearby drains.

Too many trees on a lot will produce too much shade which inhibits the growth of grass and shrubs. There are only a few shrubs which survive in densely shaded areas. However, as discussed in chapter 5, and earlier in this chapter, it is advisable to plant a reasonable number of shade trees around the house to protect against the summer's hot sun. Trees are well known as natural air conditioners because they absorb the sun's heat in the summer, and prevent it from entering the house causing overheating. In the winter after the leaves have shed they allow the warmth of the sun to enter the house. Special attention should be made to the planting location of coniferous trees. They provide shade, and also reduce the winter winds which may positively impact on winter heating costs.

The Backyard Development:

It is the back yard that requires the most care and design time because it is this part of the lot that provides the privacy, the individual/family outdoor environments, and the other practical area, the garden. On the landscape plan have these areas allocated to accommodate your family's needs. Have each member of the family write down on a piece of paper their requirements for the back yard. Some needs will be the same, and others might be too costly, but this should provide sufficient information for the family to determine the level of priority for price, placement and design of the backyard.

Deck, patio or terrace areas should be open and located off the family's living areas, dinette and kitchen. To protect against cold winds and unwanted onlookers, this area may be screened off from view with a lattice fence or shrubs, and partially roofed for protection for barbecuing and food service. These areas require good drainage away from the house with level areas for the placement of patio furniture. They can be protected against the growth of unwanted weeds by placing a polyethylene vapor barrier under the structure to inhibit their growth, which will allow for more leisure time with less maintenance activity.

Service and holding areas should be located close to the rear entrance, kitchen or garage. These areas should have sufficient space to hold items such as firewood, scrap lumber, and gardening supplies or equipment. The covered, holding area for the garbage cans should be easily accessible from the back or front yard, but hidden from sight and away from the main activity areas of the backyard.

Children's play areas should be located close to the house so that the adults can keep an eye on the children when playing in this area. It should also be located fairly close to a door in case quick access in or out is required. A low wood or wire fence may help to isolate this area, and keep the children away from gardens or other unsafe areas. A shade tree located in close proximity will provide a cool and protected area to play in. Play areas should be designed with safety in mind by eliminating any dangerous playground equipment.

Garden and lawn areas may have many different design requirements depending on the family's needs and lifestyles. These areas may consist of lawns, flower beds, or vegetable gardens. In order to obtain most benefit from these areas the backyard should be divided in two or three parts separated by changes of grade levels, sidewalks, fences, or well placed trees and shrubs. This segmentation will provide greater interest and individuality for each of the areas rather than leaving a single area cut up with small or scattered planting beds. These are often hard to maintain, and become obstacles when cutting the grass areas between. Simpler designs will provide a more effective and more organized impact on the backyard landscaping. Extra topsoil and peatmoss should be mixed and placed in the vegetable garden area with a small area set aside for composting. Compost material can be mixed into the garden in the fall to provide more natural nutrients to the soil for spring planting. A raised boarder of brick or treated wood should be used to keep the soil in these raised areas from washing onto the grass or sidewalk areas. A fence will keep unwanted animals from wandering into the garden during the growing and harvesting seasons.

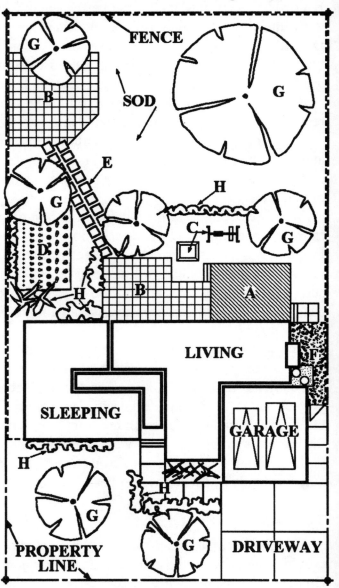

Figure 29-3. Illustration of landscaped lot.
A - Raise Deck. B - Concrete Patio.
C - Children's Play Area. D -Vegetable Garden.
E - Stone Walkway. F - Dog/Trash Holding Area.
G - Large Shade Trees. H - Low Shrubs or Hedges.

Seeding or Sodding:

Budget, time, experience and need for a completed landscaped yard will determine the question of whether to seed or sod. In either instance the rough grade must be prepared so that when the black dirt is placed over it any water will drain away from the house foundation. All shrubs, trees, flower-beds and vegetable gardens should be in place, and sectioned off with boarders, fences, sidewalks and shrubs before the black dirt is spread. The levels of these areas must also be considered before placing the black dirt and sod. Should these areas be raised or even with the sod, and will they be crossed with the lawn mower when the grass is cut?

Seeding a yard should be done between June and September, or as soon as the spring frost has left the ground, and the ground can be worked.

The subgrade or finished rough grade should be loosened up with a power cultivator, then at least 4 inches of black dirt spread evenly over the surface, and a commercial fertilizer spread on top. These materials are mixed together by raking the surface to a fine and even grade first in one direction, and then at right angles to the original direction. Before spreading the seed, use a hand roller to sufficiently compress the dirt to a level and even grade without having any raised or sunken areas, and then rake the yard lightly so the seed will fall into the loose grooves. When purchasing the seed buy a mixture that has been known to grow successfully in the area, and is sold by the local greenhouse. It is to your benefit to purchase the better quality seed. In order to properly spread the seed, rent or purchase a push-type seed and fertilizer spreader to ensure an even spread over the areas without variations in the density of the grass growth.

Rake the seeded area gently, once again in two directions, so that the seeds and the fertilizer will mix properly, and encourage the grass seed to germinate more quickly. If the soil is dry enough, use the hand roller to compress the seeds into the top soil. Then lightly sprinkle the seeded area with water to a depth of about 1 inch. Keep the seeded area moist until the grass has reached a height of not less than 2 inches, at which time the first cutting is done. When the lawn has been established, after about one month, give it a good watering only when the soil becomes dry, and then soak it thoroughly to a depth of 4 inches. Fertilize the lawn with a quality commercial fertilizer in the spring and fall to ensure a green and healthy lawn.

Sodding is usually done if there is an urgent need for a quick lawn. This method is more expensive than seeding, but it can be completed and ready for watering in a weekend.

The area to be sodded is prepared in the same way as for seeding. Not less than 4 inches of black dirt should be used over the rough grade. Fertilizer is mixed with the dirt in the same manner, but in a 25% proportion of the total amount. The remainder of the fertilizer is applied as follows: 50 percent after the black dirt has been rolled, leveled, and raked prior to laying the sod, and the remaining 25 percent spread over the snow in the winter, or in the late fall with the final watering. Sprinkle the black dirt and fertilizer lightly with water before laying the sod. Purchase the sod from a local sod farm or greenhouse, and check it for weeds, mushrooms, and other fungi. Each sod strip should be 1 to 1 1/2 inches thick, and evenly cut with the grass length being 1 1/2 to 2 inches long. When laying the sod, always stagger the joints with the next row. Newly laid sod should be hand rolled to a point where the surface becomes uniform.

After laying the sod, water the grass immediately to ensure a 5 inch penetration of moisture through the sod into the black dirt which allows the fertilizer to mix well into the black dirt under the sod. Then allow the grass to dry, but not to a point where the subsoil is dry to the touch. Keep the sodded area moist until the grass has reached a height of 3 to 4 inches at which time the first cutting can be done. Cut it initially to the 2 inch level, and after 2 weeks of several waterings when the grass roots have had time to grow, cut the grass again to the desired level. Continue watering only when the soil becomes dry, and then soak it thoroughly to a depth of 4 inches. Fertilize the lawn with a quality commercial fertilizer in the spring and fall to ensure a green and healthy lawn.

Vegetable gardens are usually small in area, but can provide an interesting and productive hobby for family members. Locate gardens on the sunny side of the yard to obtain full exposure of afternoon and evening sun. Make sure that a vegetable garden has at least 6 inches of soil composed of black dirt, sand, and peat moss. The wet compost collected during the year from house waste should be added in the fall after all the vegetables have been picked. This will add rich nutrients to the soil for the next year's growing season.

Flower gardens are decorative components of landscaping, and add color to the visual streetscape of the house. Color becomes more appealing when similar varieties and colors are united rather than scattered throughout the flower bed. Plant flowers where they can be viewed and enjoyed through a window on the inside as well as from the outside of the house.

When selecting flowers from the local greenhouse for placement adjacent to walkways, terraces, or decks, choose five or six types of hardy perennial plants which flower at different times, and plant them in groups of 8 to a dozen. Spring bulbs like tulips and gladioli can be added in random groupings between the perennials. Annual plants such as petunias or snapdragons will add color to the flower bed until it has developed and matured, and will bloom throughout the summer until the first frost.

Choose flowering plants and shrubs that have a high success and growth rate for North American gardens. Consult your local landscaper, greenhouse, and University or Government Department of Horticulture for suggestions for plants suitable to your specific area.

NOTES

CHAPTER 30

• CONTRACTS

If you are building your own home this author suggests a simple, easy to understand contract that can be used for all the suppliers and subcontractors. Attach a copy of the contract to the estimate received from the subcontractors. Be sure to review the estimates, and account for all the labor and materials suggested in this book. Read the chapters that refer directly to the suppliers or subcontractors estimates, and make sure that everything has been included. If not, add an addendum to the estimate sheet, and have the owner of the company sign it. Make sure that somewhere in the estimate or addendum it states that they are to supply all materials and labor necessary to complete the contract in a good and workman-like manner.

If the sample contract plus the subcontractor's estimate is not sufficient for yourself or the bank, standard contracts can be purchased at most stationery stores.

SAMPLE -   SUPPLIER / SUBCONTRACTOR AGREEMENT

DATE:_____

SUPPLIER/SUBCONTRACTOR

Name: _____   Estimate Date:_____

Address:_____   Total Estimate: $_____

_____   Workmans Comp. No.:_____

                    Postal Code:_____   Expiration Date:_____

Business Phone: _____ - _____ .   Tax ID Number:_____

Cellular Phone: _____ - _____ .

Guaranty or Warranty:_____

JOB ADDRESS/ LOCATION:

LOT: _____   BLOCK: _____   PLAN:_____

LEGAL ADDRESS: _____   _____

PAYMENT SCHEDULE:

Upon Completion of:_____   Amount: $_____

Upon Completion of:_____   Amount: $_____

Final Payment Due:_____   Amount: $_____

Completion Date:_____   _____ .

Lien Holdback of _____ % to be held for 45 days after satisfactory completion.

◆   Additional materials/labor to be included in estimate by Supplier/Subcontractor with no extra charges.

_____

_____

_____**or - See Attached Schedule.**

CONDITIONS:

All work to be completed in a good and workman like manner, in strict accordance with Local Building Regulations and Specifications contained in the Working Blueprints.

_____   _____
Signature of Supplier/Subcontractor   Signature of Builder/Owner

NOTES

CHAPTER 31

● CONSTRUCTION SCHEDULE

The standard start-to-completion contractor's construction schedule is usually 49 to 60 working days depending upon the weather. As a Lay Builder, and out of consideration for your lender, a minimum 65 to 80 working day construction schedule will allow for bad weather and other construction delays. When sitting down to review the schedule, the most important part will be the advance phone call made to the suppliers and subcontractors confirming their on-site start date. These calls are made to jog their memory, and ensure that their tendency to overbook will not lead to a delay in your work schedule. It will be your job to act as a subtle irritant, and remind them by phone on several occasions, and several days in advance of their start date in your house construction schedule.

The number of workers on a subtrade crew will vary the scheduling, i.e., my framer has a five man crew with some men working with him for many years. His crew work together as a unit with every man knowing what is to be done, and what is expected from the boss. Each worker has a designated job to do with minimum supervision and without error. This allows the boss to coordinate the layout and framing of the walls and floors without excessive delays due to questions, and on-site miscommunication between inexperienced workers. This framing crew will complete a 1,200 to a 1,600 square foot house from the floor joists to roof preparation stage in about 6 working days, give or take one day depending on weather or crew size on any given day.

Some items which will extend the construction schedule are:

1) Angled walls, bay windows, bow windows, and different framing methods will add 1 day to the framing schedule.

2) Several different roof slopes, valleys, and peaks that require additional stick framing will add 1 day to the framing schedule.

3) For every 200 square feet over the basic 1,600 square feet add 1 day to the framing schedule.

4) For each worker less than a five man crew add 1 working day to the framing schedule. A framing crew of less than three men will take about 12 working days.

5) If building in cold weather or before/after the building season your schedule will have to adjust.

Construction-sites left unattended as when the subtrades are not on the job site, or during the evening/night hours will experience material thefts and subsequent delays. To avoid replacement/time delays place lumber in an area that is well lit and not easily accessible. Other materials such as wire mesh, nails, caulking, and polyethylene vapor barrier delivered too early should be stored in a safe place, eg., take the small items with you in your vehicle, or have the subcontractor using these materials store them in his truck for the evening. If this is not possible, hide them where they cannot be easily seen or accessed by vehicles such as the basement, or chain heavy items such as wire mesh to a post or a section of the house with a key lock. Do not accept early delivery for items such as tubs, toilets, spas, sinks, showers, etc. If the subdivision is busy with construction crews they will very likely be stolen. Instruct suppliers to deliver these items <u>only</u> on the day of installation.

 Note: Construction insurance covers some of the potential theft or breakage of materials during construction. However, sometimes the deductible to be paid out will not cover the full cost of the item, therefore this amount will have to be paid out of pocket.

The following comprehensive, 49 day construction schedule is provided as a broad example. However, as a lay builder set your own schedule depending on the weather, the amount of time you will be able to spend on the site and make those necessary phone calls, and the number of approved construction courses attended.

◆ CONSTRUCTION SCHEDULE:

Construction Day and Phone Preparation

6 Days Prior to Construction:

- Read construction schedule,. The start date of each subtrade or supplier is marked with a *.
- Pick-up and pay for city building permit. The application should have been made 15 working days before.
- Pick-up mortgage approval and inspection sheet from the Bank. Confirm Interim Financing.
- Contact Insurance Company for liability, theft, construction, and fire insurance.
- Phone subcontractors to see if it is necessary to apply for credit for 30 - 60 day account credit.

- Order material and confirm delivery date for the concrete, lumber, windows, stairs, sand, fireplace and trusses, etc. with the actual delivery date to be confirmed 1-3 days prior to scheduled delivery.

- Confirm each scheduled start date with the framer, plumber, electrician, heating contractor, electrical trencher, water/sewer installers, surveyor, weeping tile subcontractor, garbage removal company and the tool rental company.

5 Days Prior to Construction:

- Make application for lines of credit with the selected suppliers and subcontractors.
- Meet with the City Inspector and Developer representative at the site to inspect the sidewalk, curb, and property for damage.
- Contact 1) excavator 2) cribber 3) surveyor 4) engineer to set the dates to be on-site.
- Call supplier for prices for portable toilet.

- Confirm order with sales representatives for precast steps, garage door, security, mirror/glass. Actual delivery date to be confirmed 3 to 7 days prior to the scheduled delivery.

4 Days Prior to Construction:

- Contact window supplier to get 3 copies of the window and door rough openings for the framer.
- Contact truss supplier for 3 copies of the roof truss layout for the framer, and the Engineering specifications for the City and the Bank.
- Contact the heating contractor for 3 copies of the duct layout and BTU loss calculation for the City and Bank.
- Make application at the Utility Company for the Gas Line trenching and Meter connection.
  (Allow 3 weeks for processing).

- Obtain prices from the cribber and window manufacturer for wood window bucks. Choose the supplier, and set a scheduled delivery date to the job site.

3 Days Prior to Construction:

- Contact electrician to set date for the temporary electrical service to be installed at the site on day 10.
- Contact a neighbor to see if he will allow the purchase of his power if problems arise during the construction period.
- Meet with the framer to review the plans.
- Contact excavator for on-site start time for day 1of the construction.

- Contact light fixture, appliance, and flooring suppliers to confirm meeting date to select product. Request a Saturday meeting. Set tentative delivery dates for for 35, 35, and 36 respectively.

Construction Day and Phone Preparation

2 Days Prior to Construction:

- After checking with the excavator, contact the engineer to inform them when the excavation hole will be ready for their soil tests on day 1.
- Call the lumber supplier for material delivery for the footing forms, wood pegs, reinforcing rods, and foundation ladders for day 1 at 8:00 AM.
  Also delivery of the materials for the wood foundation if applicable on AM day 2.
- Contact surveyor to confirm stakeout of the lot in late afternoon on the day prior to day 1.
- Go to the Bank to sign interim financing.
- Contact cribber for the start time to form the footings for the afternoon of day 1. Meet with cribber to review the blueprints, and confirm what he has to do on day 1.
- For winter ground heat confirm the delivery of straw or coal for the late afternoon of day 1. The cribber is to be there to spread and light the material.
- If constructing a wood foundation call the gravel supplier. The washed gravel for the foundation base should be delivered and spread late in the afternoon of day 1.

1 Day Prior to Construction:

- Contact concrete supplier for delivery time for day 2, and confirm schedule with cribber for day 2.
- Contact plumber to install the underfooting water, sewer, and sump lines for late afternoon of day 1, and confirm the schedule with the cribber.
- Confirm with the excavator as to his start time for digging hole on day 1. Inform the cribber of the schedule.
* - Surveyor on-site to stake excavation for foundation.

Day 1 of Construction:

* - Excavator at job site (7:30 - 9:00 AM) - Allow about six hours for excavation of 1,000 to 1,300 sq. ft house. Show excavator where to place soil for future perimeter fill for house settling. Have him keep the soil away from the garage, front entrance, and sewer, water, and electrical areas to be trenched.
* - Delivery at site of material for footing forms, reinforcing rods, pegs, and foundation ladder.
* - Engineer on-site to inspect the excavation soil to determine the footing size, and make soil tests for the City Soil Report. Confirm footing size on-site for cribber.
- Late afternoon delivery of washed gravel for wood foundation if applicable. Wood delivery on AM day 2.
- Delivery of straw/coal if winter frost heating is required.

- Contact the water and sewer line installers and schedule them for the PM of day 2.

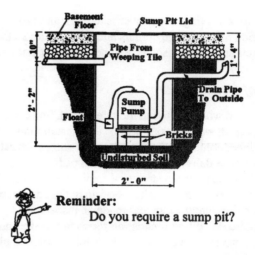

Reminder:
Do you require a sump pit?

- Phone the window supplier re: delivery of the window bucks for the morning of day 2 if they are to supply.

- Call and confirm the engineer and lumber yard for arrival times on the job site day 1.

- Post copies of the Building Permit and the Bank's Mortgage sign on the job site. They must be visible to the inspector from the street.

- Order the portable toilet from the supplier to be delivered for day 5. (Locate it out of the way of any work or material drop-off areas).

- Call weeping tile, basement dampproofing contractors to set site schedule for day 6.

- Call framer for treated basement walls for day 2, and floor joists for day 6.

- Call sewer/water line installers to confirm that they will be requesting their own inspections from the City Inspectors prior to backfilling.

- Order the concrete for the footings and pads to be delivered day 2. Check with cribber's schedule for time.

Construction Day

Day 1 of Construction continued:

* * - Cribber to form for footings and pads for basement structural posts.
* * - Plumber to dig under the footing forms for water/ sewer pipes.
* - Prior to leaving, the cribber is to mark and peg the location of the garage corners and garage piles so the water and sewer installer will not trench in that location.

Day 2 of Construction:

* - Treated wood for foundation delivered. Framer to frame basement walls for day 5 floor joist installation.
* - Cribber completes forming footings and structural pads.
* * - Concrete delivery for footings and pads.
* - Cribber to pour the structural footings and pads.
* * - Water/sewer line installers on-site in the afternoon. (Confirm that the City has made an inspection)
* - Cribber to deliver foundation forms to job site and set them against the excavation hole for securing in place on day 3.
* - Delivery of the basement window bucks on-site.

Day 3 of Construction:

* - Cribber to strip footings and structural pad forms.
* - Cribber to complete delivery of the foundation wall forms.
* - Cribber to start setting up foundation wall forms between 10:00 AM and 12:00 noon. This will allow time for the concrete footings to cure. The window bucks and foundation ladders will also be set in place.
* - If the cribber is to set the floor joists, schedule day 4 for setting the floor joists and day 5 for pouring the concrete for the foundation walls.
* - Surveyor on-site to confirm the initial foundation locations. If there are changes, request that a copy of the new plot plan be made available for pickup by day 8.

Day 4 of Construction:

* - If cribber is to install the joist package, the lumber and steel teleposts will be delivered this morning (8:00 AM).
* - If the framer installing joists, the concrete pump should be on-site 1 hour prior to pouring the foundation, and the concrete trucks arrival.
* - Delivery of the foundation wall concrete. Order at 7:00 for 8:00 delivery today. Better yet, have the cribber order the concrete. (Check with the cribber early in the morning about the weather conditions and the ordering of the concrete).
* - Mark the location of portable toilet for day 5 delivery.

Preparation

- Call window, roof truss, cabinet, exterior cladding, insulator, drywaller, roofer, soffit, fascia, finisher, painter, precast stairs, concrete finisher, etc., to inform them of the schedule. Make sure that they will be ready with 2 or 3 days notice.

- Call surveyor to confirm that the foundation location is correct for a Certificate on day 3.

- Call weeping tile and dampproofing installers to book for day 6 application and installation.

- Call lumber supplier to deliver the floor joists and steel teleposts on day 4 (8:00 AM) if the cribber is to set them, otherwise wait for (8:00 AM) day 6.

- If the framer is setting the joists order the concrete and concrete pump for day 4. The concrete pump is to be the first on the job site for set-up.

- Call framer to start framing the floor joists for day 6.

- Call the framer to confirm that the floor joist material will be delivered in the morning on day 6. Confirm that he will be ready to start.

- Contact the surveyor's office and have them Fax a copy of the plot plan to your lawyer. Notify him that one will be sent for his records.

Construction Day

Day 5 of Construction:

- If cribber set up joist package day 4, then concrete pump and concrete will be scheduled for today. See notes for day 3 order.
- If concrete poured day 4, cribber on-site stripping the foundation forms starting at about 10:00 AM. Make sure that the concrete has properly cured before stripping.
- Cribber to break off all protruding snap ties from the concrete foundation wall.
- Mark location of sump pit for weeping tile installer.
- Portable toilet to be delivered today.
- Phone City/Mortgage Inspectors. Request that an inspection be done prior to 12:00 noon backfilling on day 7.

Day 6 of Construction:

- If concrete poured day 5, strip and clean forms today.
- Lumber yard delivery of floor joist package on-site for framer 8:00 AM. Have the framer review the packing slip to make sure that all materials have been sent. Reject any materials that are warped, split, or poor quality.
* - Framer on job site setting posts, beams, floor joists, subfloor glue, and subfloor. To start late afternoon if cribber has done joist package.
* - Weeping tile / dampproofing / sump pit subtrades on site to apply and install. (Wood or conc. foundation).
- Phone to schedule the pile auger subcontractor for day 8, 1:00 PM. The cribber should pour the piles right after the pile auger is finished.

Day 7 of Construction:

- Framing crew finishing subfloor installation/glueing.
* - City and Mortgage Inspectors on-site prior to backfill. (The inspection report will be left with the framer or on the sign with the Building Permit and Mortgage ID).
- Excavator on-site at 12:00 noon for the backfilling and rough grading of the foundation and lot.
- Main floor framing package dropped off at the site. Have framer reject any inferior quality materials. Locate the package where it will not interfere with the grade beam and pile marking, or the excavator's rough grading not yet finished.
- Contact Bank Loans Officer to confirm that the inspection has been completed. Request your first Mortgage draw, and ask when it can be picked up at the lawyers.

Preparation

- If framer setting joists, call lumber supplier to confirm delivery of steel posts and floor joist package for first thing on day 6.

- Confirm with weeping tile and dampproofing subtrades for day 6 installation and application.

- Call excavator about backfilling and rough grading of site for day 7 (about 12:00 noon). Confirm time with framer. Will the site be clear of any lumber?

- Call lumber supplier about delivering the main floor framing package in the late PM of day 7. Mark location for package so that it will not interfere with backfilling, trenching, or the garage pile work areas.

- Call electrical trenching subtrade, and request day 8 AM installation. (Confirm that they will request the inspection by the City).

- Contact cribber to request that the piles and grade beam location be marked with pegs for day 8 at 11:30 AM for the 1:00 PM drilling and 3:30 PM pour.

- Call the concrete supplier to schedule pouring the piles for day 8 - delivery at 3:30 PM.

- Call roof truss supplier to confirm delivery for the afternoon of day 10.

- Check with the framer to see if he needs a crane for day 10 to lift the trusses onto the roof, or if he will be doing it by hand. Confirm who will call to order the crane.

- Pick-up Insurance Policy, and have them Fax a copy of the policy to the lawyer for the bank.

- Call electrician for the main panel installation day 10.

- Call surveyor to recheck full foundation location for Real Property Report on day 9.

Construction Day

Day 8 *of Construction:*

- Framer on-site framing the main floor walls.
- Cribber to be on-site at 11:30 AM to peg piles/garage.
* - Electrical trenching contractor on-site to install power
  to the house. (Check to make sure that they have re-
  quested an inspection of their installation by the City).
* - Pile auger contractor on-site drilling the piles 1:00 PM.
- Concrete trucks on-site at 3:30 PM. Cribber to place
  the reinforcing rods before pouring the concrete.
- Cribber placing the forms for the grade beam pour.
  Make sure that the frost void forms are placed at the
  bottom of the grade beam.
- Pick-up the new plot plan from the surveyor. Request
  a Real Property Report for PM day 9.
  (Confirm with cribber that grade beam will be formed)
- Request the cribber to order the grade beam concrete
  for 10:00 AM day 9 or to fit his schedule.
- Have the framer call the roof truss supplier to confirm
  the delivery date for afternoon of day 10.

Day 9 *of Construction:*

- Framer on-site framing main floor walls, and starting
  second story floor joists.
- Cribber to drill dowel holes and install the reinforcing
  where house foundation and grade beam wall connect.
  They will also install the grade beam framing ladder.
- Delivery of the grade beam concrete 10:00 AM.
- Surveyor on-site confirming house foundation location
  on the lot. They will provide a Real Property Report.
* - Roofer checking site. Usually they drive around checking the schedules with the framers. Confirm roofing
  material can be delivered for 12:00 day 12 when framer
  and the plumber are finished with the roof.

Day 10 *of Construction:*

- Framer on-site installing upper floor subfloor and walls.
* - Stair supplier on-site to deliver the stairs for the framer
  to install.
* - Truss supplier (and crane) on-site to deliver the roof
  trusses. Have the framer count the trusses to make
  sure they are all accounted for. Give framer a copy of
  the truss layout.
- Take plot plan and Real Property Report to lawyer's
  office, sign the Mortgage documents, and pick up the
  first mortgage draw which was requested day 7.
* - Electrician on-site to install the main power panel.
- Cribber on-site to strip the grade beam forms.

Preparation

- Call all subtrades, i.e., trusses, windows, roofer,
  etc. Inform them of the schedule.

- Call plumber to deliver the tubs and showers for
  PM day 11, and install roof stacks late PM day 12.
  (Check with framer for accurate schedule times)

- Call heating contractor to deliver the sheet metal
  roof /vent collars day 11, and mark vent locations.

- Call window supplier to deliver the windows PM
  day 12 after the framer has completed the roof.

- Contact concrete supplier to confirm cribber's time
  for day 9 delivery for grade beam.

- Contact stair manufacturer. Order the stairs for AM
  of day 10. (Get stair measurements from framer).

- Call electrician to confirm installation main power
  panel for day 10. Set electrical rough-in for day 14.

- Call surveyor late PM to find out when they will have
  the Real Property Report ready for pick-up. Copies of
  plot plan and Real Property Report may be Faxed to
  the lawyer.

- Call the plumber to arrange schedule for the basement rough-in, and installation of the roof vents
  by 12:00 Noon day 12 before roofer starts.

- Call Gas Utility Company for meter installation on
  day 20. (You will reconfirm).

- Call the fireplace supplier for installation of the fireplace for the morning of day 12.

- Call the surveyor to confirm the pick-up of the Real
  Property Report today.

- Measure the house for precast steps. Order them
  for delivery and installation on day 21. If prefab units
  will not be used, the concrete finisher will make them
  when he pours sidewalk and driveway.

Construction Day

Day 11 of construction:

- Framer on-site framing upper floor walls, trusses and roof sheathing. (Make sure that insulation stops are placed between all the trusses).
* - Heating contractor to drop off the sheet metal for the roof furnace flue, fireplace flue, and vents. He is also to mark the location of the exterior cook top/dryer vent units, all the interior floor vents, all return air grills, the furnace fresh air intake and the flue fire stops for the framer to cut to size. Have them inspect the framing to see that it will meet their requirements, and measure the chimney chase drain cap if you have one.
- Plumber to deliver the tub and shower units for the framer to install that day.

 Note: Leave protective cardboard intact so that they will not be damaged by workers.

Day 12 of Construction:

- Framer finishing the roof sheathing, fireplace framing, deck wall header, and any final framing required inside the house. Before the windows and doors arrive, have them apply the caulking around the rough openings of the window and door frames on top of the building paper where the frames will attach to the plywood.
* - Delivery and installation of the fireplace. Have them show the framer where any air intake, venting cuts, or special framing will be required before installation. (Have the installers measure for the chimney chase drain cap if you have one).
- Do a visual inspection of the house with the framer looking for any missing intake or exhaust vent cuts or holes that will be required by other subcontractors.
- Plumber on-site to install waterproof membranes and roof stacks, and rough-in for the basement, main, and upper floor pipes. (The installation of the hot water tank is to be done after the basement floor is installed).
* - Delivery of windows to the site. Before installing have the framer review the packing slip to ensure they are all there, and unbroken. (Take cranks off and remove the screens. Store them in a safe place).
- Delivery of roof materials. Have it placed directly on the roof to eliminate any theft. The roofer will start to install the roofing paper.
- Before the framer leaves, have the drywaller do a visual inspection to check blocking, backing, and insulation stops, and fix any potential framing problems.
- Purchase some cheap door locks and install them.

Preparation

- Call the sand supplier for the garage and basement to be delivered for day 15.

- Call bank to see if the interim financing cheques are ready for pick up.

- Contact the heating contractor. Update heating duct rough-in schedule for the afternoon of day 13.

- Confirm with the electrician that rough-in will be required for day 14.

- Call the drywaller. Tell them the house will be ready for insulation on day 17. Request a pre-inspection by the drywaller on day 12 before the framer completes.

- Call the window well supplier for installation day 14. (They are usually the weeping tile contractors).

- Call the soffit and fascia suppliers for installation on day 15. Gutters to be installed on PM of day 19.

- Call stucco or siding contractors for installation to start on day 18.

- Call concrete finisher to confirm readiness to drill holes in the garage grade beam for dowels, and spread and tamp sand in the garage/basement on day 16. A morning concrete pour for the basement can be done on day 17, and a morning pour for the garage can be done on day 18.

- Check the size of the garbage pile, and contact the garbage removal subtrade for a day 13 pick-up. (Have them place any large materials that can be used by other subtrades on a neat pile nearby).

- Contact the security, vacuum, and intercom suppliers for rough-ins on day 15.

Construction Day

Day 12 of Construction continued:

- Have several copies of the interim house key made. Give them to the people you trust, and leave one in a hiding place around the house just in case someone needs to gain entrance when you are not there. As this happens get the key back, and move the hiding place.

Day 13 of construction:

- Plumber on-site to complete rough-in. Have him install a common tap in the basement for the subtrades to use during construction. Do a visual inspection of the drilling, cutting, and notching that was done, and have them seal any holes with caulking, insulation stops, or vapor /air seals of any type.
- Roofers on-site to finish installing the paper, and start installing the roofing material.
- Heating contractor on-site to start the furnace duct rough-in. The furnaces will not be installed until the basement floor has been completed. In winter the furnaces are usually hung by the floor joists until the basement floor is finished.
- Pick-up interim finance cheques from bank.
* - Garbage subtrades on-site to sweep the house and remove all the garbage inside and out. Have them set aside any long lengths of lumber and large sheets of plywood. Pile them neatly near a back entrance.

 Note: Spend some time collecting some more estimates for carpets, painting, finisher, drapery, appliances, etc.

Day 14 of Construction:

- Heating contractor on-site to complete furnace duct rough-in.
- Window well subtrade on-site to install metal wells. (They will also place sufficient washed rocks around the vertical well drain connected to the weeping tile).
- Roofer on-site finishing roofing material installation.
* - Electrician on-site to start the main floor rough-in. Review the electrical blueprints for all plug, switch, and light locations. Confirm the basement rough-in for AM day 24.

 Note: If building when the weather is cold, arrange to rent a propane tank and blower to heat the house and speed up curing of the concrete floor being poured day 16.

Day 15 of Construction:

- Electricians on-site to complete rough-in today.
* - Security, vacuum and intercom installers on-site to do their rough-in.

Preparation

- Call the Utility company to confirm the gas line and meter installation on day 20.

- Call all subcontractors to inform them of your progress and their start schedule, i.e., drywaller, painter, masonry, cabinets, carpets, soffits, fascia, exterior cladding, finisher, etc. Does their schedule and yours still work?

- Confirm installation of the furnace units with the heating contractor on day 19.

- Inform plumber that the meter and gas line is to be installed on day 20.

- Call sand supplier to confirm delivery of the garage and basement sand for PM of day 15.

- Call City and Mortgage Inspectors for a completed rough-in inspection on afternoon of day 15 prior to insulation and vapor barrier installation.

- Call concrete finisher to confirm sand spreading and drilling the grade beam holes to accept the reinforcing rods on day 16. The pour for the floors

Construction Day

Day 15 of Construction continued:

* * - Soffit/fascia suppliers on-site to install same. Confirm gutters are booked for installation PM on day 19.
* - City/Mortgage Inspectors on-site for rough-in inspection prior to insulation and vapor barrier installation.
* * - Sand delivery for basement and garage. They usually use a conveyor belt to place the sand. Remove a few basement windows so they will not get broken. Place some poly or plywood over the window wells to catch any overflow sand. This will stop the sand from ending up in the weeping tile and sump, possibly clogging and burning out the sump pump.

Day 16 of Construction:

* - Soffit/fascia installers on-site to complete work. Confirm they will be on-site to install the gutters PM day 19.
* * - Concrete finisher on-site drilling holes in grade beam to accept reinforcing rods, spreading/tamping garage and basement sand, and spreading the 6 mill poly. vapor barrier. Confirm concrete delivery for the basement floor for day 17, and the garage floor for day 18. They are to confirm the times with you or the concrete supplier.
* - Pick-up #2 Mortgage draw from the lawyer if needed. Make sure that he does a title check for property liens.

 Note: Have the concrete finisher leave an excess of 12" of poly around the perimeter of the basement. This can then extend up from the basement floor, and above the frost wall for a continuous vapor barrier.

Day 17 of Construction:

* - Concrete finisher on-site to pour the basement floor.
* - Concrete trucks on-site at 8:00 AM for basement pour.
* * - Drywaller on-site to place insulation and poly. vapor barrier. Discuss with the installer about hard to get areas, drill holes, and gaps around window and door rough openings. Confirm delivery of the drywall material on-site for AM on day 19.
* - On the exterior walls mark the locations for the brick and exterior cladding. Then contact the exterior cladding supplier to confirm installation for AM of day 18, and the brick supplier for PM of day 20.

 Note: The City Building Inspector will require an inspection prior to the drywall boarding. Request an inspection for AM of day 19. The Mortgage Inspector does not require an inspection until the primer painting of the walls is completed.

Day 18 of construction:

* - Concrete trucks on-site to pour the garage floor.
* - Concrete finisher on-site to place garage floor.
* - Drywaller on-site to complete installation of the insulation and vapor barrier.

Preparation

are set for days 17 and 18.

- Pick-up some extra 6 mill poly vapor barrier from the lumber yard for garage and basement floors.

- Inform the lawyer that an inspection has been completed. Request a second Mortgage Draw. Sometimes the draw monies at this stage of construction are not needed until the end of the month. Have the lawyer inform the bank not to draw this advance until requested to save on interest charges.

- Call exterior cladding supplier to confirm installation for day 18, and set masonry (brick) for day 20.

- Call concrete supplier for delivery of the basement concrete for 8:00 AM day 17, and confirm that time with concrete finisher.

- Call drywall subcontractor to confirm schedule for installation of insulation and poly. vapor barrier day 17, and drywall delivery for AM day 19.

- Call the heating subcontractor to confirm the installation of the furnaces for AM of day 19.

- Call the framing contractor to confirm frost wall installation for day 20.

- Call concrete supplier for delivery of the garage floor concrete for 8:00 AM day 18.

- Confirm with concrete finisher re: pouring the garage floor 8:00 AM on day 18.

- Call the lumber yard to deliver the material for the frost wall to be installed by the framer at 8:00 AM day 20.

Construction Day	Preparation

Day 18 of Construction continued:

* - Exterior cladding supplier on-site placing the building felt or tyvec wrap over the exterior plywood. Inspect that they have not covered any wall plugs or wall light boxes. They will also be installing the stucco wire if applicable.

* - Utility company inspector on-site to make sure the area they will be trenching is clear of any debris. If clear he will give the approval for the installation for day 20 by leaving a date card on the site.

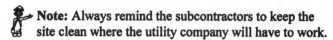 Note: Always remind the subcontractors to keep the site clean where the utility company will have to work.

Day 19 of Construction:

- City Inspector on-site to do an inspection prior to the drywall board being installed.
* - Exterior cladding installers on-site. Siding should be completed by day 20.. Stucco wire to be finished today, and first mud coat to be applied day 20.
- Heating subcontractor on-site to install the furnace and garage space heater if applicable.
* - Cabinet maker on-site to take final measurements.
- Eaves trough, gutter, and downspout installer on-site. Takes usually 4 to 8 hours depending on house size.
- Drywall scheduled for AM delivery, and boarders on-site to start. Confirm with them anticipated schedule for completion. Usually about 1 1/2 to 2 days with clean-up.

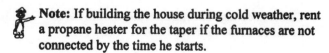 Note: If building the house during cold weather, rent a propane heater for the taper if the furnaces are not connected by the time he starts.

Day 20 of construction:

- Lumber for frost walls being delivered at 8:00 AM.
- Framer on-site to frame basement frost walls. Make sure that he wraps the vapor barrier under the wall plates for a continuous vapor barrier up the wall.
- Utility company on-site trenching the gas line, and installing the meter to the house wall.
* - Masonry subcontractor on-site to install the brick.
- Exterior cladding installer on-site to complete siding today. For stucco, the first mud coat will be applied with the final coat applied day 21, curing permitting.
- Drywall boarders on-site to finish boarding. Taper to start day 21. Before they leave, make sure that they sweep the house floor, and clean out any debris from the floor heating ducts and fresh air returns.

Preparation

- Inform framer that you have scheduled the lumber for the frost walls for day 20.

- Arrange appointment with the cabinet maker at the job site to confirm cabinet measurements for day 19.

- Call painter, finisher, floor covering, plumber, bobcat, electrician, etc., and inform them of your progress. Schedule their installation/start dates if you can.

- Call installer for the gutters, and confirm a late PM installation for day 19.

- Call Utility company to confirm line and meter installation for day 20. Make sure the location of the meter is marked with spray paint. The siding or the first coat of stucco is applied before meter installation.

- Call drywall subtrade to schedule wall taping for day 21, and the ceiling texture application on AM day 25.

- Call plumber to schedule the installation of the hot water tank, and interior gas and water lines for day 21.

- Call bobcat subtrade to do driveway and sidewalk cuts and leveling, and property grading for slope to drain for PM of day 21. Book for final grading day 41.

- Call the sand supplier and schedule delivery of sand for the driveway/sidewalks for 8:00 AM on day 22.

- Inform plumber that gas line and meter have been installed. Confirm with them that the interior gas lines are to be installed in the AM of day 21.

- If pre-cast steps ordered, contact the supplier for a PM installation for day 21.

- Call concrete finisher to confirm AM of day 22 to form for sidewalk and driveway, spread/tamp sand, drill grade beam and foundation holes to accept reinforcing rods, and cut and place wire mesh for a PM concrete pour. Set stair forming for AM day 23.

- Call parging (stucco) subtrade to apply parging to the exposed perimeter foundation on day 25.

Construction Day

Day 20 of construction continued:

Day 21 of Construction:

- Stucco subcontractor on-site to apply the final coat of stucco. If weather is cold or rainy - tarps and heat might be required to dry the stucco and parging. This could cost extra, and will add about 3 days to the schedule.
* - Bobcat on-site doing the driveway and sidewalk cuts. Have him place sufficient dirt fill under the site of the exterior stairs to allow for foundation soil to settle.
- Plumbers on-site to install the interior gas lines, flexible hoses to the furnace, and the gas line to the hot water tank. They are all usually installed and pressure tested the same day. Be sure that they request an inspection.
- Taper on-site to apply 1st and 2nd coat of joint filler. It will take about 3 days to complete with evening heat maintained at about 75° - 80° F.
* - Precast stairs on-site to be installed if applicable.

Note: Remember to call a rental company for a commercial vacuum to be ready for PM pick-up on day 25. This will be used to clean the floors, walls ducts, and window sills of drywall dust.

Day 22 of construction:

- Taper on-site for sanding and applying the 3rd and final coat of joint filler.
- Sand supplier on-site to deliver driveway and sidewalk sand. Have them place it so it will be easy to spread, and it will not interfere with the wood forms.
- Concrete finisher on-site to set forms, drill for dowels, and set reinforcing rods and wire mesh in place for driveway and walkway. If the finisher is to make the front steps, they can be formed and poured on PM of day 23.
- Concrete trucks on-site to pour the driveway and sidewalk.
- Lumber delivery of forms and pegs for driveway and sidewalks.

Day 23 of Construction:

- Taper on-site to complete final sanding, and any minor filling and sanding that is required. Do a walk-through inspection with the taper before he leaves.
- Concrete finisher on-site at 8:00 AM forming steps.
- Concrete trucks on-site at 2:00 PM to pour steps. Expect to be invoiced for a partial load, unless they waive it.

Preparation

- Call drywaller to set PM of day 25 to insulate and install the poly vapor barrier to the perimeter frost wall, floor joists, and cantilevers in the basement.

- Call electrician to schedule the basement rough-in wiring for AM of day 24.

- Schedule painter for application of exterior paint to frames/wood decor, and interior coat of primer on PM of day 24.

- Call suppliers to schedule sand/concrete deliveries for the driveway and sidewalk for day 22. Also confirm this schedule with the concrete finisher.

- Call all suppliers and subtrades who are working at the site on the 22nd to use the back entrance.

- Call lumber supplier to deliver driveway and sidewalk forms and pegs for 8:00 AM on day 22.

- Call the spindle and railing supplier to the site to measure house on day 24.

- Call cabinet supplier, and confirm a late afternoon delivery for day 26.

- Call finishing carpenter, and schedule him to install the plywood overlay for the tile and lino areas AM on day 26, and the doors, trim, and baseboards on AM of day 29. Baseboards in tile and lino areas to be cut, but not installed.

- Call concrete supplier to deliver concrete 2:00 PM the for stairs on PM of day 23. Confirm time with concrete finisher. Request a waiver of a partial load charge.

- Call drywaller to confirm AM installation of the basement insulation and vapor barrier, and ceiling texture application on day 25.

- Call Mortgage company to inspect on PM day 25.

- Call lumber yard to order finishing package and deliver late PM day 28.

Construction Day   **Preparation**

Day 23 of Construction continued:

- Call the garage door company for day 28 installation.

- Confirm basement rough-in with electrician for AM day 24.

Day 24 of Construction:

- Electrician on-site 8:00 AM for the basement rough-in.
* - Painters on-site to apply the primer paint coat to interior of house and exterior wood. They will also do the baseboards and trim.
* - Spindle and railing supplier on-site to measure house.

- Call drywaller to book installation of attic insulation for AM of day 28.

- Call spindle and railing supplier to book installation for 8:00 AM of day 28.

Day 25 of Construction:

- Drywaller on-site to install insulation and vapor barrier in basement.
- Concrete finisher on-site to strip step forms only. Driveway and sidewalk forms should be left until landscaping.
- Drywall subtrade on-site to spray ceiling texture.
- Stucco installer on-site to apply parging around the perimeter foundation of the house.
* - Pick-up the commercial vacuum from rental company. Vacuum the house from top to bottom including the heating ducts and furnaces. This should take about 4 to 6 hours, so do it late afternoon and evening.
- Mortgage Inspector on-site for inspection # 3, after completion of primer paint coat and texture.
- Contact the appliance supplier and the plumber for sink and cook top templates to be picked up on day 26 so that the cabinet installer and finisher can cut the cabinets to accept them.
- If possible, return the vacuum unit to reduce charges to an hourly rate instead of a full day.

- Call finishing carpenter to confirm installation of plywood overlay 8:00 AM day 26.

- Call the lumber yard to deliver plywood overlay for 8:00 AM on day 26. Have them place it in an area of the house that will be carpeted.

- Call cabinet supplier, and confirm a late afternoon day 26 delivery. To start installation on day 27.

- Inform lawyer that an inspection has been completed. Request the 2nd and 3rd draws to pay the current bills.

- Call concrete finisher to book the late afternoon of day 27 to apply sealer to driveway, steps, garage pad, and walkways.

- Call heating subcontractor to book him to do the site measuring on day 27 for the ceiling fan vent installation before the attic insulation is placed.

Day 26 of Construction:

- Pick up templates from appliance supplier and plumber.
- Lumber supplier on-site to deliver the plywood overlay.
- If not already done, return the commercial vacuum to the rental company before the 24 hour rental time is up.
* - Finishing carpenter on-site to install the plywood overlay over clean subfloor prior to the cabinets being delivered. If possible have him use **glue and nails**.
* - Kitchen cabinets being delivered at site late afternoon, and to be installed day 27.

- Call and confirm schedules with drywaller for attic insulation, garage door supplier for installation, and spindle supplier for installation on day 28.

- Book finishing carpenter for installation of finishing the package on AM of day 29.

- Call lumber supplier to confirm the delivery of the finishing package for PM day 28 at back door.

Note: By this time phone calls will be coming on a daily basis from the suppliers and subcontractors re: payment. Tell them that once the Mortgage draw is available they will be paid. Be sure you get a good discount for cash payments.

Construction Day

Day 27 of Construction:

- Concrete finisher on-site to apply the concrete sealer to the driveway, walkway, and garage pad. Have him place a barrier on the sidewalk to keep people from walking on the wet sealer, and request all workers to enter through the back door for the next few days.
- Cabinet installer on-site. Before he starts to assemble the cabinets, inspect them for scratches, dints, marks, or undesirable discoloration, and reject them if present.
- Give the templates from the plumber and appliance supplier to the cabinet installer to fit the sinks and cook-top. Measure the rough openings for the refrigerator, oven, microwave, and compactor to make sure they are the correct size, and that the appliances will fit into the cabinets.
- Heating contractor on-site to measure the fan duct sizes, and make sure that all fan and exhaust vents are installed, insulated, and sealed. Book him for AM of day 29 to mark the duct size in the cabinet for the finisher.
- Pick up 2ⁿᵈ and 3ʳᵈ mortgage draws from the lawyer.

Day 28 of Construction:

- Cabinet supplier on-site to install the cabinets.
- Spindle and railing supplier on-site to install same.
- Attic insulation supplier on-site to install the loose fill R 40 cellulose insulation.
- * - Garage door supplier on-site to install the garage door and garage door openers.
- Painter on-site to finish exterior wood/window frames, and prime the spindles and railings. Make sure that the heat is registering about 75° F. for the paint to dry.
- Lumber supplier on-site to deliver the finishing package. Take the door knobs, towel bars, bathroom fixtures, or any other items, and store them in your trunk to pull out when needed. Have him do a material count.

Day 29 of Construction:

- Cabinet suppliers final day on-site to install counter tops and cabinets. Make sure that he has cut out the rough openings to fit the sinks and cook top.
- Heating subcontractor on-site to mark the down vent size and location for the finisher. When cut he will install the ducting for the cook top.
- Finisher on-site to install doors, casing, fireplace mantel, and the baseboards in the carpeted areas. Have him install the door hardware and locking

Preparation

- Call drywaller to confirm the installation of the attic insulation for day 28.

- Call framer to set schedule to install the deck on day 30.

- Call telephone company to book installation of the phone, and a new phone number. Allow them about two weeks. Set a tentative date to confirm on day 35 for day 37 installation.

- Call the rental company to book a jumping tamper, not a vibrator for day 29.

- Call spindle supplier to confirm 8:00 AM installation of railings, etc., day 28.

- Call finishing carpenter to confirm the installation of all interior finishing materials for 8:00 AM on day 29.

- Call painter and book one of his workers to finish painting all exterior window and door frames, and apply a primer to railings for late PM on day 28.

- Call the lumber supplier to have the deck package delivered for the framer at 8:00 AM on day 30.

- Call the painter to schedule days 32 and 33 for the final interior painting of the walls, doors, and any other areas to be painted.

- Call the plumber to schedule installation of all the fixtures and sinks for 8:00 AM on day 39.

- Call the electrician to schedule the delivery and installation of the light fixtures, switches, and plugs on day 35.

- Call the appliance supplier to schedule the delivery of the built-in appliances at 8:00 AM on day 35 for the electrician to install. Do not have the refrigerator, micro-wave or other appliances delivered as they can be stolen. Wait for day 42.

- Call the vacuum, security, and intercom supplier to complete their installation on day 35.

Construction Day   **Preparation**

Day 29 of Construction continued:

devices a couple of days before move in (day 41) to reduce scratches or breakage. Baseboards for tile/lino areas cut, but not installed. Should finish tomorrow.
- Pick-up tamper from the rental company and tamp all around the house perimeter especially in areas where there was electrical, water, and sewer trenching.

- Call lumber supplier to confirm the delivery of the deck package for 8:00 AM on day 30.

- Call the framer to confirm the delivery of the deck package for installation on day 30.

Day 30 of construction:

- Finishing carpenter on-site to complete the installation of the doors, trim, closet rods, shelves, bathroom wall fixtures, medicine cabinets and any other required materials. Schedule him to install the door knobs, dead bolts and door stops on day 41.
- Lumber supplier on-site to deliver the deck package.
- Framer on-site to install the deck package. Have him review the packing slip, and reject any bad materials. Make sure that a 6 mill poly vapor barrier is placed on the ground under the deck to keep weeds from growing.

- Call the carpet suppliers, and schedule the carpet, lino, and tile measuring for day 32.

- Call the supplier for the mirrors and shower doors to schedule their measuring on day 32.

- Pick-up sufficient poly vapor barrier from the lumber yard for the framer to place under the deck. So they will not rip it, have them install it after they finish installing the joists.

Day 31 of Construction:

- Finishing carpenter on-site to complete the interior door and trim package. Confirm that the hardware installation is set for day 41.
- Framing crew on-site to finish the deck and railing installation.
- It is always a good idea to have the heating ducts and furnaces vacuumed before you move in. Get some price quotations and schedule the cleaning for day 41.

- Call the carpet, lino, and tile suppliers to set the installation dates for days 36 through 39.

- Call the bobcat subtrade to do the final grade, and slope to drain on day 41 for final survey on day 42.

- Call the heating contractor to schedule the installation of the fresh air and floor grills on day 39.

Day 32 of Construction:

* - Mirror and shower door supplier on-site to measure the sizes and locations. Schedule their installation for day 39.
* - Carpet, lino, and tile supplier on-site to do final measurements. Confirm schedule of the lino for day 36, the tile for days 36, 37 and 38, and the carpet for day 39.
- Painter on-site to complete all the interior and exterior painting. They should return on the day before move-in to do any final paint touch ups.

- Call the curtain and venetian suppliers to meet at the house on day 33 to measure and provide an estimate. Get at least three price quotations.

- Call the electrician to confirm fixture delivery and installation on day 35.

- Call furnace cleaners for price quotations, and set date for PM day 41.

Day 33 of Construction:

- Painter on-site to complete all the interior and exterior painting. Should be finished today.
* - Curtain and venetian suppliers on-site to measure and provide an estimate.

- Call the appliance, vacuum, security, and intercom suppliers to confirm their scheduled installations and delivery for day 35.

Construction Day

Day 34 of construction:

- Prior to all the suppliers and subcontractors finishing their contact, do a walk-through with them, and closely inspect everything that might require more work, or touch-ups. Remember - you are the one that has to live in the house, or if you plan on selling it, to sell a house that you are proud of.

 Note: Even if you are planning to have a cleaning staff come in before move in day, it would be a good idea to rent a commercial vacuum cleaner to vacuum the basement, closets, cabinets, ducts, and garage areas for drywall dust and sawdust. Book it for day 40 before the furnace cleaner is in the house on day 41.

Day 35 of Construction:

* - Appliance supplier on-site to deliver the appliances.
- Vacuum, security, and intercom suppliers on-site to do the final installation and testing.
- Electricians on-site to complete the installation and testing of fixtures, plugs, and switches.
- Before you leave set the security.

Day 36 of Construction:

- Electricians on-site to complete fixture installation.
- Lino and tile suppliers on-site to start installing. The lino will be completed today.
- Before you leave set the security.

Day 37 of Construction:

- Tile subtrade on-site to continue installing the tile. Installer to grout all tile tomorrow.
* - Telephone company on-site to install phone line.
- Before you leave set the security.

Day 38 of construction:

- Tile subtrade on-site to finish and grout all tile areas.
- Drywall taper on-site to do touch-ups on walls. Make sure the touch-up areas are sanded.
- Before you leave set the security.

Preparation

- Call the window covering supplier that you have chosen to schedule day 42 for installation.

- Call the floor covering suppliers to confirm the start of the lino installation for day 36.

- Call telephone company to confirm phone number installation and test for day 37.

- Call the plumber to confirm the fixture installation for AM of day 39.

- Call the insurance company to convert your construction policy to a home owners policy for day 42.

- Call drywaller, and schedule day 38 for final touch-ups, if needed, by the taper.

- Call floor covering supplier to confirm carpet installation on day 39.

- Call heating subcontractor to schedule final test, grill installation, and servicing for PM on day 39.

- Call several maid services for prices, and schedule cleaning for PM on day 42.

- Call painter to book for final touch-ups for day 42.

- Call portable toilet company to pick-up unit day 40.

- Call rental company to rent a commercial vacuum cleaner for day 40.

- Call and confirm schedule with plumber, carpet, and heating subcontractors for day 39.

- Call and confirm schedule with finishing carpenter to install door knobs, dead bolts, and door stops on day 41.

Construction Day   **Preparation**

Day 39 of Construction:

- Plumber on-site to complete final installations, servicing, and tests.
- Carpet subtrade on-site to install the underlay and carpet. Should be finished today.
- Heating subtrade on-site to complete any final installations, servicing, and tests.
- Mirror/shower door supplier on-site to install and clean.
- Before you leave set the security.

- Call surveyor to schedule a survey for a final grading certificate on AM of day 42.

- Call appliance supplier to schedule the delivery of the refrigerator, micro-wave, and laundry washer and dryer on AM of day 42.

- If you have a water line to the refrigerator, call the plumber to have it connected late AM on day 42.

- Call maid service to confirm cleaning of house day 42 prior to move-in.

Day 40 of Construction:

- Portable toilet supplier on-site to remove unit.
- Go to Registration Office or Motor Association to have the address changed on your driver's licence.
- Complete all the necessary phone calls and letters for your address and phone number change. Go to the Post Office to transfer your mail.
- Pick-up the commercial vacuum cleaner at the rental company. Vacuum the entire house up and down. Have it returned for the hourly and not the daily rate.
- Before you leave set the security.

- Call bobcat subcontractor to confirm day 41 final grading and slope to drain of your lot.

- Call the furnace cleaner, and confirm day 41 for vacuuming the ducts and furnace.

- Call the window covering supplier to confirm the installation for day 42.

- Call surveyor to confirm schedule on day 42.

Day 41 of Construction:

- Bobcat on-site to grade property for the surveyor's Grading Certificate.
* - Furnace cleaner on-site to vacuum the ducts, and clean and service the furnace.
- Finisher on-site to install door hardware, door stops, and to do any servicing.
- Before you leave set the security.

- Call appliance supplier to confirm the delivery of the refrigerator, micro-wave, and laundry washer and dryer on AM for day 42.

- If applicable, call the plumber to confirm the water line connection to the refrigerator late AM of day 42.

 Note: Before you can occupy the house you must get an occupancy permit from the City. Have the City and Mortgage Inspector do a final inspection on day 42.

Day 42 of construction:

- Appliance supplier on-site for the final delivery of the appliances.
- Plumber on-site to connect the water line to the refrigerator if applicable.
- Surveyor on-site to survey the site for a Grading Certificate.
- Painter on-site to do final paint touch-ups.
- Window covering supplier on-site to install the drapery and venetians.
* - Maid service on-site at PM to clean the entire house.

- Call the lawyer and inform him that the final inspections have been completed. Request that the final draw on the mortgage be made. Make an appointment to sign the final mortgage documents on day 43.

- Call the surveyor to find out when you will be able to pickup the Grading Certificate. When it is completed have them Fax a copy to the City Inspector, and the lawyer's office.

Day 42 of construction continued:

- City and Mortgage Inspectors on-site to make their
  final inspections. Request an occupancy permit or
  move in approval from the City Inspector.
- Before you leave set the security.

Day 43 - Move In:

- Go to the lawyer's office to pick up the final draw,
  and sign the final Mortgage documents.

- Make application for any tax rebates as soon as
  you can after paying all your invoices.

- Negotiate with the suppliers and subcontractors.
  Will they will give you a reduction if they are paid
  early or in cash?

*** MOVE IN. ***

What is left to do

Transfer:

- ☐ Bank records.
- ☐ Club memberships
- ☐ Dental records
- ☐ House and Car Insurance, etc.
- ☐ Legal contracts
- ☐ Medical, Optometrist records
- ☐ School records
- ☐ Veterinarian records

Notify:

- ☐ Church
- ☐ Credit card companies
- ☐ Friends and relatives
- ☐ Government (family allowance, etc.)
- ☐ Magazine subscriptions
- ☐ Mail order accounts
- ☐ Motor vehicle registration
- ☐ Post Office

Renew:

- ☐ Cable television
- ☐ Diaper service
- ☐ Electricity
- ☐ Food service
- ☐ Garbage collection
- ☐ Gas, fuel, or oil company
- ☐ Home cleaning service
- ☐ Laundry, dry cleaning service
- ☐ Lawn, snow removal services
- ☐ Milk, bakery deliveries
- ☐ Newspaper delivery
- ☐ Nursery, day care service
- ☐ Telephone, water, service
- ☐ Water softener service

NOTES

SECTION 5: APPENDICES

APPENDIX 1

METRIC CONVERSION CHART

* Measurements *

Length	Hard Conversion	Soft Conversion	Length	Hard Conversion	Soft Conversion
1/4"	6.0 mm	6.4 mm	10"	250.0 mm	254.0 mm
5/16"	7.0 mm	7.9 mm	12"	300.0 mm	304.8 mm
3/8"	9.5 mm	9.5 mm	1'	300.0 mm	304.8 mm
1/2"	12.0 mm	12.7 mm	2'	600.0 mm	609.6 mm
5/8"	15.5 mm	15.9 mm	4'	1200.0 mm	1219.2 mm
11/16"	16.5 mm	17.5 mm	6'	1800.0 mm	1828.8 mm
3/4"	19.0 mm	19.1 mm	8'	2400.0 mm	2438.4 mm
1"	25.0 mm	25.4 mm	10'	3000.0 mm	3048.0 mm
2"	50.0 mm	50.8 mm	12'	3600.0 mm	3657.6 mm
4"	100.0 mm	101.6 mm	14'	4200.0 mm	4267.2 mm
6"	150.0 mm	152.4 mm	16'	4800.0 mm	4876.8 mm
8"	200.0 mm	203.2 mm	18'	5400.0 mm	5486.4 mm

* Lengths *

1 mm (millimeter) = 0.3937 inch
1 m (meter)   = 3.28084 feet
1 m (meter)   = 1.09361 yards
1 km (kilometer) = 0.621371 mile

1 inch = 25.4 mm
1 foot = 0.3048 m
1 yard = 0.9144 m
1 mile = 1.60934 km

* Area *

1 square centimeter = 0.155000 sq. inch.
1 square meter   = 10.7639 sq. feet
1 square meter   = 1.19599 sq. yards
1 ha (hectare)   = 2.47105 acres

1 square inch = $6.4516 \ cm^2$ (squared)
1 square foot = $0.0929030 \ m^2$ (squared)
1 square yard = $0.836127 \ m^2$ (squared)
1 acre   = 0.414686 hectare

* Volume *

1 litre = 0.219969 gallon
1 gallon = 4.54609 litres

* Mass *

1 kilogram = 2.20462 pounds
1 pound   = 0.453592 kilograms

◆ Appendix 1 Continued

* Insulation *

R - 8 = RSI 1.4   R - 28 = RSI 4.8
R - 12 = RSI 2.1   R - 32 = RSI 5.3
R - 20 = RSI 3.2   R - 40 = RSI 7.0

* Poly. Vapor Barrier *

2 mill = millimeters thick
4 mill = millimeters thick
6 mill = millimeters thick

* Lumber Sizes *

Length	Hard Conversion	Soft Conversion	Length	Hard Conversion	Soft Conversion
1"	25 mm	25.4 mm	6"	150 mm	140.0 mm
2"	50 mm	38.0 mm	7"	175 mm	165.0 mm
2 1/2"	62 mm	51.0 mm	8"	200 mm	184.0 mm
3"	75 mm	64.0 mm	9"	225 mm	210.0 mm
3 1/2"	87 mm	76.0 mm	10"	250 mm	235.0 mm
4"	100 mm	89.0 mm	11"	275 mm	260.0 mm
4 1/2"	112 mm	102.0 mm	12"	300 mm	286.0 mm
5"	125 mm	114.0 mm	14"	350 mm	337.0 mm
5 1/2"	137 mm	127.0 mm	16"	400 mm	387.0 mm

Notes: A standard metric stud size = 38 mm x 2310 mm **or** 1 1/2" x 92 5/8" imperial. The typical ceiling height of a room is 2400 mm **or** 8 feet imperial. Structural steel posts (adjustable) = 75 mm **or** 3 inches.

* Concrete *

2000 psi = 15.0 MPa
2500 psi = 17.5 MPa
3000 psi = 20.7 MPa

* Reinforcing Rods *

2 = 5/16" = 6 mm
4 = 3/8" = 10 mm
5 = 7/16" = 12 mm

* Drywall *

3/8" = 9.5 mm
1/2" = 12.7 mm
5/8" = 15.9 mm

* Door Sizes *

All residential doors are standard 6' - 8" in height = 2032 mm.

Door widths: 2' - 0" = 609 mm
                   2' - 4" = 711 mm
                   2' - 6" = 762 mm
                   2' - 8" = 812 mm
                   3' - 0" = 914 mm

GLOSSARY OF TERMS
****

ADHESIVE: A substance capable of holding material together by surface attachment. A general term that includes glue, cement, mastic, and paste.

AGGREGATE: Materials such as sand, rock, and gravel used to make concrete.

AIR DRIED: Wood seasoned by exposure to atmosphere in the open or under cover, without artificial heat.

AIR-ENTRAINED CONCRETE: Concrete that has air in the form of minute bubbles mechanically mixed into the concrete to provide structural strength and quicker curing.

ANCHOR BOLTS: Bolts embedded in concrete used to hold a structural member in place.

ATTIC OR ROOF SPACE: The space between the top floor ceiling and roof.

BACKFILL: The replacement of earth after excavation.

BALUSTER: Turned spindle-like, vertical stair member which supports the stair railing.

BALUSTRADE: A railing consisting of a series of balusters resting on a base, usually the treads, which supports a continuous stair or handrail.

BASEBOARD: A molded board placed against a wall around a room next to the floor to conceal the joint between the finished floor and wall finish.

BASEMENT: The base story of a house, usually below grade.

BATTEN: A strip of wood placed across a surface to cover the joints. Usually found at floor or wall stud separations.

BAY WINDOW: A rectangular, curved, or polygonal window, or group of windows usually supported on a foundation extending beyond the main wall of the building.

BEAM: A principal structural member used between posts, columns, or walls to support vertical loads.

BEARING PARTITION: A partition which supports a vertical load in addition to its own weight.

BEARING WALL: A wall which supports a vertical load in addition to its own weight.

BEVEL: To cut to an angle other than a right angle, such as the edge of a board or door.

BID: An offer to supply, at a specific price, materials, supplies, and equipment, or the entire structure or sections of a structure.

BLEMISH: Any defect, scar, or mark that tends to detract from the appearance of wood or a finished surface.

BOARD: Lumber less than 2 inches thick.

BOARD FOOT: The equivalent of a board 1 foot square and 1 inch thick.

BRICK CONSTRUCTION: A type of construction in which the exterior walls are bearing walls made of brick.

BRICK MOLDING: A molding for windows and exterior door frames. Serves as the boundary molding for brick or other siding material, and forms a rabbet for the screens, and/or storm sash or combination door.

BUTT: Type of door hinge. One leaf is fitted into the space routed into the door frame jamb, and the other into the edge of the door.

CABINET: Case or box-like assembly consisting of shelves, doors, and drawers used primarily for storage.

CABINET DRAWER GUIDE: A wood strip used to guide the drawer as it slides in and out of its opening.

CABINET DRAWER KICKER: Wood cabinet member placed immediately above, and generally at the center of a drawer to prevent tilting down when pulled out.

CASEMENT: A window in which the sash swings on its vertical edge, so it may be swung in or out.

CASING: The trimming around a door or window, either outside or inside, or the finished lumber around a post or beam.

CAULK: To seal and make waterproof cracks and joints, especially around window and exterior door frames.

CLOSET POLE/ROD: A round wood or metal molding installed in clothes closets to accommodate clothes hangers.

COLUMN: Upright supporting member circular, square, or rectangular in shape.

CORNER BEAD: Molding used to protect corners. Also a metal or plastic reinforcement placed on corners before plastering.

COUNTERFLASHING: Flashing used on chimneys at the roof-line to cover shingle flashing, and to prevent moisture entry.

DEAD LOAD: The weight of permanent, stationary construction included in a building.

DECAY: Disintegration of wood substance due to action of wood destroying fungi.

DIMENSION LUMBER: Lumber 2 to 5 inches thick, and up to 12 inches wide.

DOOR FRAME: An assembly of wood or metal parts that form an enclosure and support for a door. Door frames are classified as interior or exterior.

DOOR STOP: A spring stopper screwed to the face of a baseboard or door frame to prevent the door from swinging through.

DORMER: A projecting structure built out from a sloping roof. Usually includes one or more windows.

DOUBLE GLAZING: Two panes of glass in a door or window, with an air space between the panes. They may be sealed hermetically as a single unit, or each pane installed separately in the door or window sash.

DRIP CAP: A molding which directs water away from a structure to prevent seepage under the exterior facing material. Applied mainly over window and exterior door frames.

DRY ROT: A term loosely applied to many types of decay, but especially to that which, when in an advanced stage, permits the wood to be easily crushed to a dry powder.

DRYWALL: Pre-manufactured materials used for wall covering which do not need to be mixed with water before application.

EAVES: The margin or lower part of a roof that projects over an exterior wall. Also called the overhang.

EXPANSION JOINT: A bituminous fiber strip used to separate units of concrete to prevent cracking due to dimensional change caused by shrinkage and variation in temperature.

FASCIA: A wood member used for the outer face of a box cornice where it is nailed to the ends of rafters and lookouts.

FLASHING: Sheet metal or other material used in roofing and wall construction (especially around chimneys and vents) to prevent rain or other water from entering.

FLOOR AREA: The gross floor area, less the area of the partitions, columns, and stairs and other openings.

FLUE: The space or passage in a chimney through which hot smoke, gas, or fumes rise. Each passage is called a flue, which with the surrounding material, makes up the chimney.

FLUSH: Adjacent surfaces even, or in same plane (with reference to two structural pieces).

FOOTING: The spreading course or courses at the base or bottom of a foundation wall or column.

FOUNDATION: The supporting portion of a structure below the first-floor construction, or grade, including the footings, which transfers the weight of the building load to the ground.

FRAMING: The timber structure of a building which gives it shape and strength including interior and exterior walls, floor, roof and ceilings.

FURRING: Narrow strips of wood spaced to form a nailing base for another surface. Furring is used to level, and form an air space between the two surfaces to give a thicker appearance to the base surface.

GABLE: That portion of a wall contained between the slopes of a double-sloped roof, or that portion contained between the slope of a single-sloped roof and a line projected horizontally through the lowest elevation of the roof construction.

GLAZING: The process of installing glass into sash and doors. Also refers to glass panes inserted in various types of frames.

GROUT: A thin mortar used in masonry and tile work.

GUTTER OR EAVES TROUGH: Wood or metal trough attached to the edge of a roof or eaves to collect and conduct water from rain and melting snow away from the roof.

HEADER: Horizontal structural member that supports the load over an opening, such as a window or door. Also called a lintel.

HEADROOM: The clear space between floor line and ceiling contained in a stairway.

HIP ROOF: A roof which rises from all four sides of a building to meet at the roof peak.

HOLLOW CORE DOOR: Flush or molded door with a core assembly of strips of wood, plastic, or units which support the outer faces.

HOSE BIB: A water faucet mounted on a wall that is threaded so a hose connection can be attached.

INTERIOR TRIM: General term for all the molding, casing, baseboard, and other trim items applied within the building by the finishing carpenters.

INSULATION: (Thermal) Any material high in resistance to heat transmission that is placed in structures to reduce the rate of heat loss.

JACK RAFTER: A short rafter framing between the wall plate and a hip rafter, or a hip or valley rafter and ridge board.

JAMB: The top and two sides of a door or window frame which contact the door or sash; top jamb and side jambs.

JOINT CEMENT: A powder which is mixed with water and applied to the joints between sheets of gypsum wallboard or

JOIST: One of a series of parallel framing members used to support floor and ceiling loads, and supported in turn by other beams, girders, or bearing walls.

JOIST HANGER: A steel section shaped like a saddle, and bent so it can be fastened to a beam or structural member to provide end support for joists, headers, trusses, etc.

KILN DRIED: Wood seasoned in a mechanical kiln by means of artificial heat, with controlled humidity and air circulation.

LAZY SUSAN: A circular revolving cabinet shelf used in corner kitchen cabinet units.

LEADER OR DOWNSPOUT: A vertical pipe that carries rainwater from the gutter to the ground or drain.

LINEAL FOOT: Having length only, pertaining to a line one foot long -- as distinguished from a square foot or cubic foot.

LINTEL: A horizontal structural member supporting the load over an opening such as a door or window.

LIVE LOAD: The total of all moving and variable loads that may be placed on a structure of a building.

LOOKOUT: Structural member running between the lower end of a rafter and the outside wall. Used on the underside of the roof sheathing at the overhang to support the soffit and fascia.

LUMBER: The product of the saw and planing mill not further manufactured than the sawing stage.

MASONRY: Stone, brick, hollow tile, concrete block, or tile, and sometimes poured concrete, gypsum blocks, or other similar materials, or a combination of same, bonded together with mortar to form a wall, ledge, buttress, etc.

MECHANICAL EQUIPMENT: In architectural and engineering practice. All equipment included under the general heading of plumbing, heating, air conditioning, gasfitting, and electrical work.

MESH: Expansed metal or woven wire used as a reinforcement for concrete, plaster, or stucco.

MILLWORK: The term used to describe products which are primarily manufactured from lumber in a planing mill or wood-working plant including moldings, door frames and entrances, blinds and shutters, sash and window units, doors, stair-work. kitchen cabinets, mantels, cabinets and porch work.

MOISTURE CONTENT: The amount of water contained in wood. Expressed as a percentage of the weight of even-dry wood.

MOLDING: A relatively narrow strip of wood usually shaped to a curved profile throughout its length. Used to accent and emphasize the ornamentation of a structure, and to conceal surface or angle joints.

MORTAR: A substance produced from prescribed proportions of cementing agents, aggregates, and water which gradually sets hard after mixing.

NONBEARING PARTITION: A partition extending from floor to ceiling which supports no load other than its own weight.

NOSING: The part of a stair tread which projects over the riser, or any similar projection. A term applied to the rounded edge of a board.

ON CENTER: A method of indicating the spacing of framing members by stating the measurement from the center of one member to the center of the succeeding one.

PARGING: Thin coat of plaster applied to stone, concrete, or brick to form a smooth or decorative surface.

PARTITION: A wall that subdivides space within any story of a building.

PARTY WALL: A wall used jointly by two parties under easement agreement, and erected at or upon a line separating two parcels of land that may be held under different ownership.

PILE: A heavy timber, or pillar of metal or concrete which is forced into the earth or cast in place to form a structural foundation member.

PITCH: Different variations of inclines or slopes, as of roofs or stairs. Rise divided by the span.

PITCHED ROOF: A roof which has one or more angled surfaces sloping at an angle greater than that required for drainage.

PLAN: A drawing representing any one of the floors or horizontal cross sections of a building, or the horizontal plane of any other object or area.

PLASTER: A mixture of lime, cement, and sand used to cover outside and inside wall surfaces.

PLUMB: Exactly perpendicular or vertical; at right angles to the horizon or floor.

PLUMBING STACK: A general term for the vertical main of a system of soil, waste, or vent piping.

PREFABRICATED CONSTRUCTION: Type of construction so designed as to involve a minimum of work at the site. Usually comprising a series of large units manufactured in a plant.

PRESERVATIVE: Substance that will prevent the development and action of wood-destroying fungi, borers of various kinds, and other harmful insects that deteriorate wood.

RABBET: A rectangular shape consisting of two surfaces cut along the edge or end of a board.

RADIANT HEATING: A method of heating usually consisting of coils, pipes, or electric heating elements placed in the floor, wall, or ceiling.

RAFTER: One of a series of structural members of a roof designed to support roof loads. The rafters of a flat roof are sometimes called roof joists.

RAIL: Cross or horizontal members of the framework of a sash, door, blind, or other assembly.

RAKE: The trim members that run parallel to the roof slope, and form the finish between the roof and wall at a gable end.

RELATIVE HUMIDITY: Ratio of amount of water vapor in air in terms of percentage to the total amount it could hold at the same temperature.

RESILIENT: The ability of a material to withstand temporary deformation with the original shape being assumed when the stresses are removed.

RETAINING WALL: Any wall subjected to lateral pressure other than wind pressure. Example -- a wall built to support a bank of earth.

RISER: The vertical stair member between two consecutive stair treads.

ROOFING: The materials applied to the structural parts of a roof to make it waterproof.

ROOF RIDGE: The horizontal line at the junction of the top edges of two roof surfaces where an external angle greater than 180 degrees is formed.

ROUGH-IN: The work of installing all pipes in the drainage system, and all water pipes to the point where connections will be made with the plumbing fixtures. Also applies to partially completed electrical wiring, and other mechanical aspects of the structure.

ROUGH LUMBER: Lumber cut to rough size with saws, but which has not been dressed or surfaced.

ROUGH OPENING: The opening formed by the framing members.

SADDLE: A small gable type roof placed on back of a chimney on a slope roof to shed the water or debris.

SASH: The framework which holds the glass in the window.

SCAFFOLD: A temporary structure and platform to support workmen and materials during construction.

SEALER: A liquid applied directly over unfinished wood or concrete surfaces for the purpose of sealing the surface from the penetration of water or other chemicals.

SHAKES: Hand or machine-split shingles made of wood.

SHEATHING: The structural covering. Consists of boards or prefabricated panels that are attached to the exterior studding or rafters of a structure.

SHEATHING PAPER: A building material used in wall, floor, and roof construction to resist air passage.

SHIM: A thin strip of wood, sometimes wedge-shaped, for plumbing or leveling wood members. Especially helpful when setting door and window frames.

SIDING: The finish cover of the outside wall of a frame building. Many different types are available.

SILL: The lowest member of the frame of a structure, usually horizontal, resting on the foundation and supporting the uprights of the frame. Also the lowest member of a window or outside door frame.

SMOKE ALARM: An electric device which sounds an alarm when sensing the presence of smoke, or air carrying combustible products relating to fire.

SOFFIT: The underside of the members of a building, such as staircases, overhangs, cornices, beams, and arches. Also called drop ceiling and furred-down ceiling.

SOIL STACK: A general term for the vertical main of a system of soil, waste, or vent piping.

SOLAR ORIENTATION: Directional placement of a structure on a building site to obtain the maximum benefits of sunlight.

SPAN: The distance between structural supports such as walls, columns, piers, beams, girders, and trusses.

SPECIFICATION: A written document stipulating the kind, quality, and sometimes the quantity of materials and workmanship required for a construction job.

SQUARE: Unit of measure -- 100 square feet -- applied to roofing material and some types of siding.

STAIR LANDING: A platform between two flights of stairs.

STAIRWAY, STAIR, OR STAIRS: A series of steps, with or without landings, or platforms, usually between two or more floors of a building.

STAIRWELL: The framed opening which receives the stairs.

STEP FLASHING: Rectangular or square pieces of flashing used at the junction of shingled roof and walls.Also called shingle flashing.

STOOP: A small porch, veranda, platform, or stairway outside an entrance to a building.

STORM DOOR: An extra outside door for protection against inclement weather.

STORY: That part of a building compressed between any floor, and the floor or roof next above.

STUD: One of a series of vertical wood or metal structural members in walls and partitions.

SUBFLOOR: Boards or panels laid directly on the floor joists over which a finished floor will be laid.

TAPING: In gypsum board or drywall construction, the masking of joints between two sheets by means of paper tape which is smoothed over with joint cement.

THERMOSTAT: An instrument that controls automatically the operation of heating or cooling devices by responding to changes in temperature.

THREE-WAY SWITCH: A switch designed to operate in conjunction with a similar switch thereby controlling one outlet from two points.

THRESHOLD: A wood, plastic, or metal member that is beveled or tapered on each side, and used to close the space between the bottom of the door, and the sill or floor underneath. Sometimes called a saddle.

TOE KICK: A recessed space at the floor line of a base kitchen cabinet or other built-in units. Permits one to stand close without striking the vertical surface with the toe.

TOENAILING: To drive a nail at a slant with the initial surface in order to permit it to penetrate into a second member.

TONGUE-AND-GROOVE LUMBER: Any lumber, such as boards or planks, machined in such a manner that there is a groove on one edge and a corresponding tongue on the other.

TOP PLATE: In construction, the horizontal member nailed to the top of the partition or wall studs.

TREAD: The horizontal part of a step on to which a foot is placed.

TRIM: The finish materials in a building, such as moldings applied around openings (window trim, door trim), or at the floor and ceiling of rooms (baseboard, cornice, picture molding).

TRUSS: A structural unit consisting of such members as beams, bars, and ties usually arranged to form triangles. Provides rigid support over wide spans with a minimum amount of material.

VALLEY: The internal angle formed by the two slopes of a roof.

VALLEY RAFTER: A rafter which forms the intersection of an internal roof angle.

VAPOR BARRIER: A watertight material used to prevent the passage of moisture or water vapor into or through structural elements (floors, walls, ceilings).

VENEERED WALL: A frame building wall with a masonry facing (example -- single brick). A veneered wall is non-load bearing.

VENT: A pipe installed to provide a flow of air to or from a drainage system, or a circulation of air within such a system to protect trap seals from siphonage and back pressure.

VENTILATION: The process of supplying and removing air by natural or mechanical means. Such air may or may not have been conditioned.

WALL PLATES: In wood frame construction, the horizontal members attached to the ends of studs. Also called top or bottom plates depending on their location.

WARP: Any variation from a true or plane surface. Warp includes bow, crook, cup, and twist, or any combination thereof.

WATER REPELLENT: A solution, primarily composed of paraffin wax and resin in mineral spirits or other man made chemicals, which penetrates wood and retards changes in its moisture content.

WATER TABLE: A ledge or slight projection in the earth's structure which carries the water away or to a given point. The level below which the ground is saturated with water.

WEATHERING: The mechanical or chemical disintegration and discoloration of the surface due to the action of dust and sand carried by winds, and the alternate shrinking and swelling of the surface fibers that come with the continual variation in moisture content brought by changes in weather. Weathering does not include decay.

WEATHERSTRIPPING: Strips of felt, rubber, metal, or other materials fixed along the edges of doors or windows to keep out drafts and reduce heat loss.

WEEPHOLE: A small hole, as in a retaining wall, to drain water to the outside. Commonly used at the lower edges of masonry cavity walls.

NOTES

_____

_____

_____

_____

_____

_____

_____

_____

_____

_____

_____

_____

_____

NOTES

◆

NOTES

NOTES

After reviewing many of the 'how to' publications, I found most were published with a focus on the technical aspects of construction with much of the information relating to only specific areas of residential construction, i.e., plumbing, heating, electrical, decks, etc. The average layman will not understand or need to understand these technical aspects of home construction.

This publication is different because it places special emphasis on presenting all the practical information related to house construction in a manner to be clearly understood by people with little or no previous building experience. Every element of the construction process is covered with examples, descriptions of the work needed, specifications, check lists, and over 127 illustrations. The book also includes a daily step-by-step construction schedule, and personal stories about the many situations that confront the self-contractor. This information is also useful for the would-be purchaser of completed homes and the book becomes a valuable resource manual. This book provides them with a detailed, question and answer process that will assist them in selecting the most suitable subdivision, house orientation and lifestyle features.

For those who do not wish to take on the full responsibility, and will have a contractor build their dream home or renovate an existing home, the daily step-by-step guide is intended to assist them to follow, understand, and supervise the contractor's work as the house is being constructed or renovated.

David Caldwell

A GREAT GIFT IDEA FOR YOUR FRIENDS
(Available at quality Bookstores and Home Improvement Centers)

I would like to take this opportunity to thank you for purchasing

"The Layman's Guide To Contracting Your Own Home".

It is our hope that the information contained in the publication will provide you, "the novice builder", the basic knowledge held by many professional home builders, and therefore make your construction project a more enjoyable and rewarding experience.

For future publications, I am always looking for funny or frustrating construction stories, cost saving products, and new methods of construction to assist the "lay builder", I would like to hear from you.

Please ☑ check appropriate box(es).

Name:_____

Street:_____

City:_____

Province/State:_____

Postal Code/Zip Code_____

☐ New Construction

☐ Existing Home Renovation

☐ Hiring a Contractor

☐ Contracting Yourself

☐ Purchasing a Home

Designs By Caldwell   **General Delivery, St.Albert, Alberta, Canada T8N 5X4**   **Phone: (403) 459-3664**